SOLAR HEATING DESIGN

SOLAR HEATING DESIGN

BY THE f-CHART METHOD

WILLIAM A. BECKMAN
Professor of Mechanical Engineering

SANFORD A. KLEIN
Postdoctoral Fellow

JOHN A. DUFFIE
Professor of Chemical Engineering

Solar Energy Laboratory
University of Wisconsin Madison

A WILEY-INTERSCIENCE PUBLICATION

JOHN WILEY & SONS, New York • Chichester • Brisbane • Toronto

Library of Congress Cataloging in Publication Data

Beckman, William A
Solar heating design, by the f-chart method.

"A Wiley-Interscience publication."
Bibliography: p.
Includes index.
1. Solar heating. 2. F CHART (Computer program)
I. Klein, Sanford A., 1950- joint author.
II. Duffie, John A., joint author. III. Title.
TH7413.B4 697'.78 77-22168
ISBN 0-471-03406-1

Printed in the United States of America

10 9 8 7 6 5 4 3 2

To Sylvia, Janet, and Pat

PREFACE

The purpose of this book is to describe a practical method for sizing solar space and water heating systems, i.e., systems which collect solar energy, store the energy, and distribute it as needed to heat a building and to heat water for domestic use. Two types of systems are considered, one based on the use of liquids as the heat transfer medium and the other based on air. These are expected to be common system configurations.

The major parts of the solar heating systems considered here are a solar collector which heats either liquid or air, an energy storage unit which may be either a water tank or a pebble bed, and an auxiliary furnace or heater. The auxiliary energy source supplies heat as needed when the collected solar energy is insufficient to meet the entire heating needs, which we refer to as the heating load.

It is technically possible to build a solar heating system which would supply 100% of the annual heating load, and which would then not require an auxiliary heater. A solar heating system designed to supply all of the energy required during the coldest months would then be greatly oversized (i.e., capable of supplying far more energy than needed) during other months of the year. We know that there is a large economic penalty resulting from oversizing a solar heating system. In almost all cases, it is more economical to design a solar heating system to supply part of the annual heating load, and provide an auxiliary energy source (conventional furnace, fireplace, wood burning stove, etc.) to supply additional energy as needed. As we will see, the economics of solar heating systems cause their design to be more critical than that of conventional systems (where oversizing results in little economic penalty).

Solar heating systems are a good example of the "law of diminishing returns," For a particular heating system, the first 20 square meters (about 200 square feet) of collector area might, for example, provide 40% of the annual heating load. Adding another 20 square meters might provide an additional 30% while the next 20 square meters would provide only another 15% of the load. So, more heat is provided per unit collector area by a small system than by a large system. Since solar collectors are expensive, the design problem is to select the size (and type) of solar collector which, in conjunction with an auxiliary furnace, will supply the entire heating load at the least possible cost. This book provides a method of solving this problem.

The problem discussed in this book is concerned with the selection of the most economical sizes of the solar collector, the storage tank, and the associated heat exchangers. We do not address the many practical problems that the designer will face in building a solar heated house. We do not discuss the sizing of air ducts to reduce noise and to insure uniform flow distribution. We do not discuss insulation requirements for the house, or the construction problems associated with installing collectors on a roof. Many of these problems are treated in standard engineering texts and in manufacturers' literature.

Each of the six chapters deal with a particular aspect of the design of solar heating systems. Each chapter is nearly complete in itself, containing introductory material, references to pertinent literature, and examples which illustrate the major points. The first four chapters present methods of estimating the various quantities that are needed in Chapter 5 to determine system thermal performance and then in Chapter 6 to determine economical designs.

Chapter 1 begins with notes on solar heated buildings which have already been constructed. Most of these systems are similar to either the liquid or the air heating systems considered in this book. The components and operation of liquid and air heating systems are described.

Chapter 2 is concerned with the thermal performance of flat-plate solar collectors. The theory which relates collector performance to its construction and to the operating conditions is presented in an abbreviated form. A more complete treatment of this subject can be found in the book, <u>Solar Energy Thermal Processes</u> by Duffie and Beckman (1974). (References are indicated in the text by authors and dates; complete details of the references are given in the bibliography at the end of the text.) The treatment given here provides an adequate understanding of collector test procedures and results which are needed to design solar heating systems.

Chapter 3 describes the effects of collector orientation on the overall performance of solar heating systems. Collector orientation affects solar heating system performance in two ways. It affects both the total amount of solar radiation striking the collector surface, and the fraction of this radiation transmitted through the transparent covers and ultimately absorbed on the collector surface. Methods of estimating the monthly solar radiation incident on the collector and the monthly average transmittance-absorptance product are presented.

The performance of a solar heating system is strongly dependent upon the size of the heating load (i.e., energy needs). It is not possible to accurately estimate long-term solar heating system performance without considering the heating load. Chapter 4 describes the degree-day method of estimating the space heating load and a method of estimating the domestic water heating load.

A method of estimating the useful output of solar heating systems is presented in Chapter 5. The procedure uses the results of collector tests, heating load calculations, and climactic data to estimate, for a particular system, the fraction of the monthly heating load supplied by solar energy. The design procedure presented here, referred to as the "f-chart" method, was developed by S. A. Klein in his Ph.D program at the University of Wisconsin. Various parts of this design procedure have appeared in other publications (Klein et al. (1976a), Klein et al. (1976b,c), Beckman et al. (1977)) which the reader may wish to consult for additional background and explanation.

Using the methods for estimating thermal performance described in Chapter 5, we show in Chapter 6 how fuel and system cost information can be used to determine the economic optimum design of a solar heating system. The costs of a solar heating system are primarily first costs, i.e., substantial investments are made in order to save on future operating costs. In the simplest terms, the building owner borrows money from the mortgage company to purchase a solar heating system which reduces his heating bill. A more complete analysis takes into account interest and principle payments to the mortgage holder, reductions in fuel costs, increases in insurance and maintenance costs, and changes in property and income taxes. The amount of fuel saved by using solar energy can be estimated using the procedure described in Chapter 5. The amount of money the building owner saves by installing a solar heating system is the difference between the cost of the fuel he saves (plus any net tax savings) and his extra payments to the mortgage company for interest and principle on the extra mortgage taken out to pay for the solar heating equipment (plus any extra expenses for maintenance, insurance, etc.). Chapter 6 describes methods for doing these calculations which account for expected inflation of fuel and other costs.

Throughout the book, there are a series of examples of the several calculations needed to determine thermal performance and economic design. These examples, taken together, present all the calculations needed to design economic solar heating systems. The reader may find it advantageous, after having studied the book for the first time, to go through the complete series of examples in order to better understand the calculations from beginning to end.

The engineer or designer who designs several systems will find that many of the calculations are repetitive, and once completed, they will apply to a variety of installations in a given location. The calculations which appear onerous at first will be considerably simplified as the reader gains experience with them. All of the calculations can be done with any inexpensive electronic calculator that can calculate exponential functions.

Many people are unfamiliar with the concept of effectiveness which is used in this text to express the performance of the heat exchangers in solar heating systems. Appendix 1 defines the concept and describes a simple method of determining the effectiveness of a heat exchanger from performance data supplied by the manufacturer. In Appendix 2, the monthly average climactic data required to estimate the long-term performance of solar heating systems are tabulated for a large number of North American locations.

Units present a problem. Much of the latest research and development in the solar energy field is reported in SI (Systems International) units, a modern version of the old metric system. However, architects, engineers, and builders primarily use English units. In view of the trend toward metrication, we have used SI units to do calculations, but have presented important results in English units as well. Two approximations allow easy translations of many quantities from one set of units to the other: A square meter is approximately ten square feet, and a kilojoule (kJ) is nearly equivalent to a BTU. A detailed set of conversion tables appears in Appendix 3.

The reader new to the field of solar energy will encounter some new terminology in this book. We have attempted to define these new terms where they first appear in the text. In addition, a glossary of terms is presented as Appendix 4.

The calculations needed to size solar heating systems are organized by worksheets; blank copies are included as Appendix 5, and the reader may wish to duplicate these to facilitate multiple calculations. These worksheets are used in the examples throughout the text to calculate solar radiation on the collector surface, heating loads, system thermal performance, and life cycle economics. Once the reader has become familiar with the use of these worksheets, the process of sizing solar heating systems will be greatly simplified.

The concepts described in this book have been incorporated into an interactive computer program, called FCHART, which is available from the Solar Energy Laboratory of the University of Wisconsin - Madison. A brief description of this program is included as Appendix 6.

We have tried to include in this book everything the reader needs (added to his engineering abilities) to size solar heating systems. At the same time, the book can be considered to be a companion volume to _Solar Energy Thermal Processes_ by Duffie and Beckman (1974). With a few exceptions, we use the same nomenclature and calculation procedures. The theory we outline in Chapters 1 and 2 of this book is a condensed version of several major chapters in _Solar Energy Thermal Processes_, and we refer the reader to it for a more complete discussion of all of the background material. The f-chart design method, which is the essence of this book, has been developed since 1974.

The research on which much of the work in this book is based was done under the sponsorship of the National Science Foundation (NSF) and the Energy Research and Development Adminstration (ERDA). The University of Wisconsin - Madison and the Wisconsin Alumni Research Foundation through the Graduate School of the University have also contributed substantially to this program.

W.A. Beckman
S.A. Klein
J.A. Duffie

June 1977
Madison, Wisconsin

CONTENTS

SOLAR HEATING DESIGN

CHAPTER 1
SOLAR HEATING SYSTEMS

1.1 NOTES ON SOLAR HEATING EXPERIMENTS

In this book, we are concerned with "active" solar heating systems, i.e., those which use equipment to collect, store, and distribute solar heat in a controlled manner. The systems we consider are those which use solar collectors as fluid heaters; the fluid is then pumped or blown to transport the energy from the collector to storage and from storage to the house. Other systems use movable insulation, rather than fluids.

The term "solar house", may also be applied to buildings in which architectural design is used to obtain solar gains in the winter (and reduce them in the summer) and so reduce heating (and cooling) loads. These "passive" solar heating systems are not treated in this book. We consider that there is no substitute for energy-conserving architectural design and our concern in this book is with methods for meeting the energy needs for space heating (and hot water) which are not eliminated by intelligent architectural design.

Solar heating is not a recent concept. Domestic solar water heaters have been in wide use for many years. A number of industries manufacture solar water heaters in Australia, Israel, Japan, the United States, and elsewhere.

The first extensive study of solar energy utilization for space heating began in 1939 at Massachusetts Institute of Technology. A series of four small structures, each partially heated by solar energy, were successively built from 1939 to 1960 (Hottel and Woertz (1942), Hesselschwert (1950), and Engebretson (1964)). The last of these, MIT Solar House IV in Lexington, Massachusetts, was designed to provide about two-thirds of the annual space and water heating loads. The performance of the heating system was carefully measured and reported for two heating seasons. The system used 60 square meters of water heating collectors for the compact 135 m^2 floor area house. The collectors were drained when not in use to prevent water from freezing in the collectors. Energy storage was provided in a 5670-liter (1500-gallon) water storage tank. After several years of experimental development, the solar heating system was removed from the building.

The Denver Solar House, designed and built by Lof in 1958, uses a solar heating system differing from MIT Solar House IV primarily in that it uses air, rather than a liquid, to transfer heat. The performance of the Denver Solar House has been reported for its first years of operation (Lof et al. (1964)) and for the 1974-1975 heating season (Ward and Lof (1975)). The 50 square meters of collector area in this system provide about 25% of the annual heating load for the building, which has a floor area of about 300 square meters (3200 square feet). This system is still in operation today, and it requires little maintenance beyond that of a conventional heating system.

Since these early efforts, numerous solar heated buildings have been constructed. A survey of most recent efforts has been compiled by Shurcliff (1977). Unfortunately, measurements of performance are available for very few of these. Although surveys show that there are a variety of possible solar heating system designs, most systems are similar in many respects. Almost all practical solar heating systems employ a flat-plate solar collector (see Chapter 2), energy storage of a capacity sufficient to supply about one day of winter heating, and an auxiliary energy source such as a furnace using conventional fuel. Most of these existing systems are similar to either the standard liquid or air heating systems described in the following sections.

1.2 LIQUID-BASED SOLAR SPACE AND WATER HEATING SYSTEMS

A schematic diagram of a typical liquid-based solar heating system is shown in Figure 1.1. This system uses liquids (generally water or an antifreeze solution) as the heat transfer fluid and water as the storage medium. Flat-plate solar collectors are used to transform incident solar radiation into thermal energy. This energy is stored in the form of sensible heat in a liquid storage tank and used as needed to supply the space and water heating loads. If the collectors are not drained at night or during periods of excessive cloudiness, an antifreeze solution is generally circulated through the collectors to avoid freezing. In this case, a liquid-to-liquid heat exchanger is used between the collectors and the tank because it is more economical than the alternative of using the antifreeze solution as the storage medium.

A water-air heat exchanger, referred to as the load heat exchanger, must be used to transfer heat from the storage tank to the building. An additional liquid-to-liquid heat exchanger is used to transfer energy from the main storage tank to a domestic hot water system. A domestic hot water system consists of a preheat tank which supplies solar heated water to a conventional water heater. A conventional furnace (i.e., auxiliary heater) is provided to supply energy for the space heating load when the energy in the storage tank is depleted. Controllers, relief valves, pumps, and pipes make up the remaining equipment.

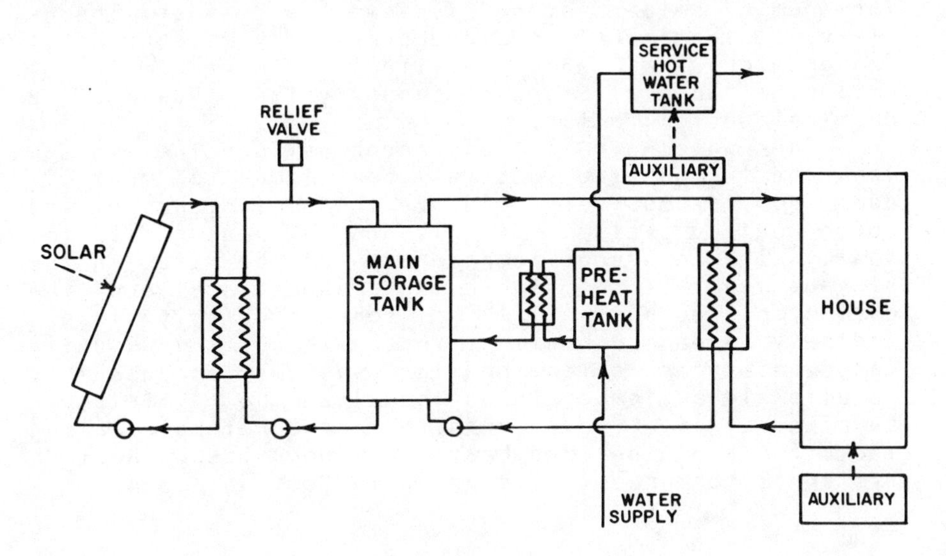

FIGURE 1.1
SCHEMATIC DIAGRAM OF A LIQUID-BASED SOLAR HEATING SYSTEM

After many computer simulations, experiments, and years of practical experience by many people in the field, a number of design recommendations for solar heating systems of this type have evolved, as indicated in Table 1.1. These recommendations should be used only as general guidelines, since manufacturers of solar heating equipment may have their own recommendations. The terminology used in Table 1.1 will be fully explained in later sections.

Usual design practice is to have the collector fluid flowrate at about 0.015 l/s per square meter of collector area (0.022 gpm per square foot of collector). Economic studies (e.g., Lof and Tybout (1973) and others) have concluded that the best storage capacity is in the range of 50 to 100 liters of stored water per square meter of collector area (about 1.25 to 2.5 gallons per square foot of collector). A house with 50 square meters (540 square feet) of collector would then require a water storage tank of approximately 3750 liters (1000 gallons). The thermal performance of solar heating systems is rather insensitive to the amount of storage capacity as long as it is greater than about 50 liters per square meter of collector area. Storage capacity is thus not a critical design factor.

The water-air heat exchanger between the storage tank and the heating load must be sized so that it does not excessively penalize the performance of the solar heating system. If this heat exchanger is too small, the average temperature of the water in the storage tank will be higher than necessary and the collector output will be correspondingly lower. Ordinary baseboard heaters, for example, are usually inadequate for solar heating systems, because they require higher temperatures than can be efficiently supplied by solar collectors. A method of calculating the effects of the load heat exchanger size on heating system performance is discussed in Section 5.3-3.

TABLE 1.1
DESIGN RECOMMENDATIONS FOR LIQUID-BASED SOLAR SPACE AND WATER HEATING SYSTEMS

Collector Flowrate (50-50 Ethylene Glycol-Water)	0.015 l/s-m^2 (0.022 gpm/ft^2)
Collector Slope and Orientation	Latitude plus 10° facing due south is best; however, differences from the optimum slope or orientation as much as 15° have little effect.
Collector Heat Exchanger	$F_R'/F_R > 0.9$ (See Section 2.4)
Storage Capacity	50 to 100 l/m^2 (1.25 to 2 gal/ft^2)
Load Heat Exchanger	$1 < \varepsilon_L C_{min}/UA < 5$ (See Section 5.3-3)
Domestic Water Preheat Storage Capacity	1.5 to 2.0 × capacity of conventional water heater

1.3 SOLAR AIR HEATING SYSTEMS

A common configuration of a solar air heating system is shown in Figure 1.2. Other arrangements of fans and dampers can be devised to result in an equivalent flow circuit. Air is heated in the flat-plate solar collector and circulated to either the house or to a pebble bed. Energy is stored in the pebble bed by heating the pebbles with the circulating hot air. At night, or in cloudy weather when the available solar energy is insufficient to meet the heating load directly, air is warmed as it is circulated through the warm pebble bed and into the house. Auxiliary energy is supplied from the furnace when the energy stored in the pebble bed is depleted. Energy required for domestic hot water is provided in some systems by heat exchange from the hot air leaving the collector to a domestic water preheat tank, as in the liquid system. The hot water is further heated, if required, by a conventional water heater. During summer operation, it is advisable not to store solar energy in the pebble bed; as a result, a manually operated storage bypass is a usual part of this design (not shown in Figure 1.2).

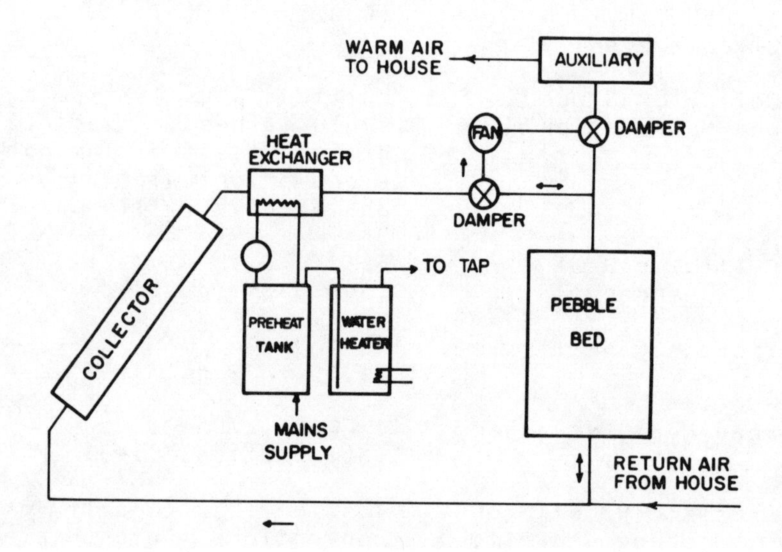

FIGURE 1.2
SCHEMATIC DIAGRAM OF A SOLAR AIR HEATING SYSTEM

Design recommendations for solar air heating systems are given in Table 1.2. When the cost of blowing air through the collectors is considered, the economic optimum air flowrate turns out to be in the range of 5 to 20 l/s per square meter of collector area (1 to 4 cfm per square foot of collector). The economic optimum storage capacity for the pebble bed is in the range of 0.15 to 0.35 cubic meter of pebbles per square meter of collector area (0.5 to 1.15 cubic foot per square foot of collector). A house with 50 square meters (540 square feet) of collector would need about 12.5 cubic meters (440 cubic feet) of pebbles. This would require a room of dimensions 2.5 m x 2.5 m x 2 m (8 ft x 8 ft x 7ft). To maximize energy storage capacity and minimize blower requirements, uniform size pebbles with a diameter ranging from 1 to 3 centimeters (about 0.5 to 1.5 inches) should be used. The exact size of the pebbles is not as critical as size uniformity. It is desirable to use washed river pebbles to reduce the dust content in the bed although crushed rock is acceptable. Air filters should be installed in the ducts on both ends of the pebble bed.

TABLE 1.2
DESIGN RECOMMENDATIONS FOR SOLAR AIR HEATING SYSTEMS

Collector Air Flowrate	5 to 20 l/s-m^2 (1 to 4 cfm/ft^2)
Collector Slope and Orientation	Latitude plus 10° facing due south is best; however, differences from the optimum slope or orientation as much as 15° have little effect.
Storage Capacity	0.15 to 0.35 m^3 of pebbles/m^2 (0.5 to 1.15 ft^3/ft^2)
Pebble Size	1 to 3 cm (0.5 to 1.5 in)
Bed Length in Flow Direction	1.25 to 2.5 m (4 to 8 ft)
Domestic Water Pre-heat Tank Capacity	1.5 to 2 x capacity of conventional water heater
Pressure Drops	
Pebble Bed	2.5 to 7.5 mm H_2O (0.1 to 0.3 in H_2O)
Collectors	5 to 20 mm H_2O (0.2 to 0.8 in H_2O)
Ductwork	1 mm H_2O/15 m (0.08 in H_2O/100 ft)
Ductwork Insulation	2.5 cm (1 in) fiberglass
Leakage	Ductwork seams must be sealed

1.4 CONTROLS FOR SOLAR HEATING SYSTEMS

Solar heating systems require controls that are different from conventional systems. First, there must be a method of turning on the collector pump or fan when the collector can supply energy. This is usually done by a differential temperature controller which measures the difference in temperature between the outlet of the collector and the bottom of the storage unit. When the collector fan or pump is off and this temperature difference rises above some fixed value, the pump or fan is turned on. When fluid is flowing and the temperature difference falls to near zero, the fan or pump is turned off.

The building also needs a control that is not conventional. The best operation of a solar energy system is obtained when the collected solar energy is used as soon as possible. The control is best arranged so that when the thermostat in the house calls for heat, the system will deliver what it can from the storage unit (or directly from the collectors in the case of air systems). If the solar energy cannot meet the load, a second stage of the thermostat turns on the auxiliary furnace which supplies the balance of the heating needs. In some designs, the furnace takes over entirely when the solar energy system can not meet the load. This may result in a simpler system, but with some penalty in thermal performance.

1.5 SUMMARY

Liquid-based solar heating systems are currently more common than the air heating systems. This is because liquids are easy to transport and they are superior on a volume basis to any other practical, sensible, heat storage material. The disadvantages of the liquid system however, are that heat exchange with the building air is required and precautions against freezing, boiling, and corrosion must be taken. This is essential because a fluid leak could cause extensive damage, a problem not encountered in an air system. (The thermal performance of liquid and air systems of similar design is neary comparable, as shown in Section 5.4.)

The reader will note in this chapter that in each of the systems we have discussed, there is both a solar heating system and a (more or less) conventional heating system. As we will see in later chapters, it is not economical to provide 100% of the heating load with solar energy, and a solar heating system should be thought to consist of two major parts, the solar energy part and the conventional part. The economic advantage of the combination is obtained when the fuel savings from using solar energy more than compensates for the increased first cost of the solar heating equipment. The major problem we address in this book is how to select the type and size of the solar heating equipment which, in combination with a conventional heating system, will supply the entire heating load at the least total cost.

CHAPTER 2
FLAT-PLATE SOLAR COLLECTORS

2.1 DESCRIPTION

The flat-plate solar collector is the basic device used in solar space and domestic water heating systems. The operation of a flat-plate solar collector is, in concept, simple. Most of the solar energy incident on the collector is absorbed by a surface which is "black" to solar radiation. Part of the absorbed energy is transferred to a circulating fluid, while the rest is lost by heat transfer to the surroundings. The heat carried away by the fluid, the useful energy gain of the collector, is then either stored or used to supply the heating load.

The essential parts of the collector are: the absorber plate, generally made of metal with a nonreflective black finish to maximize the absorption of solar radiation; pipes or ducts to circulate either liquid or air in thermal contact with the absorber plate; thermal insulation for the back side and edges of the plate; one or more air spaces separated by transparent covers to provide insulation for the top of the plate; and a housing to assure a durable and weatherable device. Section views of both liquid and air heating collectors are shown in Figure 2.1.

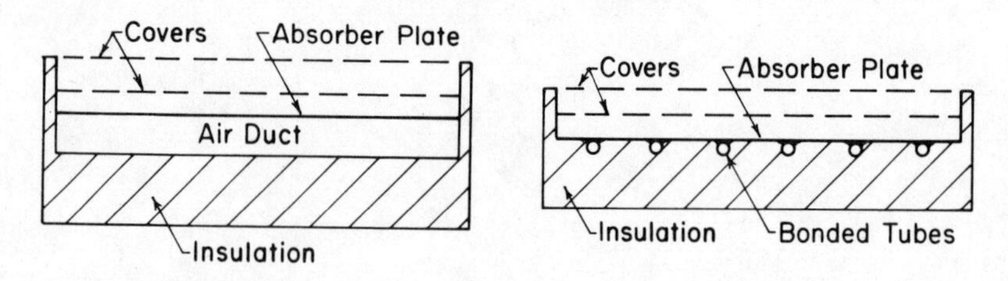

FIGURE 2.1
SCHEMATIC DIAGRAMS OF LIQUID AND AIR HEATING COLLECTORS

The transparent covers are usually made of glass. Glass has excellent weatherability and good mechanical properties; it can be very transparent if it has a low iron oxide content, and it is relatively inexpensive.

The disadvantages of glass as a cover material are that it is breakable and heavy. Plastics can also be used. They are generally less subject to breakage, light, and inexpensive in thin sheets. However, plastics generally do not have as high a resistance to weathering as glass; the surface can be scratched and many plastics degrade and yellow with age which reduces its transmittance to solar radiation and leads to loss in mechanical strength. One additional advantage of glass over plastics is that glass absorbs or reflects all of the long-wave radiation (heat) incident on it from the solar heated absorber plate, thus reducing radiation losses from the plate more effectively than plastics which transmit part of the long-wave radiation.

Flat-plate collectors absorb both beam and diffuse radiation. Beam radiation is that part of the solar radiation which casts shadows. Diffuse radiation is reflected and scattered by clouds and dust before it reaches the ground and it does not cast shadows. Flat-plate collectors are usually mounted in a stationary position on a building or other structure with an orientation dependent upon the location and the time of year in which the solar energy system is intended to operate. The effects of collector orientation on solar heating system performance are considered in Chapter 3. Flat-plate collectors provide low temperature heat which is the form of energy required for space and water heating.

Concentrating solar collectors, such as parabolic or Fresnel focusing collectors and the non-imaging compound parabolic concentrators (CPC), could be used in solar heating systems. Most concentrating collectors will use only beam solar radiation. The advantage of concentrating collectors over flat-plate collectors is that the area from which heat losses occur is reduced so that the energy can be delivered at higher temperatures than are attainable with flat-plate collectors. However, higher temperatures have little or no value for space and water heating. In the case of most focusing concentrators, the collector mounting must track the position of the sun. The non-imaging systems usually require repositioning a few times each year. In this book, we treat systems which use flat-plate collectors which are mounted in a fixed position.

2.2 COLLECTOR THEORY

In order to better understand the discussion to follow, it is necessary to distinquish between instantaneous collector performance (i.e., the performance of the collector at a given point in time, as a function of the meteorological and operating conditions occurring at that time) and the long term performance. In practical operations, a collector in a solar heating system operates over a wide range of conditions during a year. It will run hot and at low efficiency at times, and cool and at high efficiency at other times. These variations are all taken into account in the design method described in Chapter 5.

In order to determine how a collector works over the range of varying operating conditions it will encounter in solar heating, it is necessary to understand the theory relating instantaneous collector performance to meteorological and operating conditions. We will see that, for our purposes, two numbers are needed to describe collector performance, one relating to how energy is absorbed and the other to how energy is lost. These numbers are best determined from collector tests, in which instantaneous collector efficiency is measured over an appropriate range of conditions. In this section, we review the essential ideas of the theory which governs instantaneous collector performance. The material presented here is presented in greater detail in Hottel and Woertz (1942), Hottel and Whillier (1956), Bliss (1959), and in <u>Solar Energy Thermal Processes</u>.

The useful energy gain from the collector at a given time is the difference between the amount of solar energy absorbed by the absorber plate and the energy lost to the surroundings. The equation that applies to almost all practical flat-plate collector designs is:

$$Q_u = F_R A[I_T(\tau\alpha) - U_L(T_i - T_a)] \qquad 2.1$$

where

Q_u is the rate at which useful energy is collected [W,BTU/hr]

A is the collector area [m^2, ft^2]

F_R is the collector heat removal efficiency factor

I_T is the rate at which solar radiation is incident on the collector surface per unit area [W/m^2, $BTU/hr\text{-}ft^2$]

τ is the solar transmittance of the transparent covers

α is the solar absorptance of the collector plate

U_L is the collector overall energy loss coefficient [$W/C\text{-}m^2$, $BTU/hr\text{-}F\text{-}ft^2$]

T_i is the temperature of the fluid entering the collector [C,F]

T_a is the outside ambient temperature [C,F]

The terms in the collector equation are explained in more detail in the following sections.

2.2-1 SOLAR ENERGY ABSORBED BY THE COLLECTOR PLATE

The incident radiation on the collector at any time, I_T, includes three parts: beam radiation, diffuse radiation, and depending on the slope of the collector and the nature of the surroundings, some radiation reflected from the ground or surroundings. When a collector is tested, I_T is measured by a pyranometer positioned at the same inclination as the collector. (As shown in Chapter 5, measurements of I_T at frequent time intervals are not needed to estimate the long-term performance of a solar heating system once collector test results are obtained. What is needed by the f-chart method described in Chapter 5 is the monthly average radiation incident on the collector surface. The most commonly available data are monthly average radiation on a horizontal surface. A method for calculating the monthly average radiation on tilted surfaces from the horizontal surface data is discussed in Sections 3.2 and 3.3.)

The rate at which solar energy is absorbed by the collector plate at any time is the product of the incident radiation, I_T, the fraction transmitted by the cover system, τ, and the fraction absorbed by the collector surface, α. Both τ and α are functions of the materials and the angle of the incident radiation (i.e., the angle between the perpendicular to the surface and the direction of the solar radiation). The beam, diffuse, and reflected components of the incident solar radiation strike the collector surface at different angles. As a result, the transmittance and absorptance must each be calculated as a weighted average of these components. (A method of estimating the monthly average product of the collector transmittance and absorptance, needed for the design procedure in Chapter 5, is discussed in Sections 3.4 and 3.5.)

2.2-2 THERMAL LOSSES FROM THE COLLECTOR

Losses occur from the collector by several mechanisms. Heat is lost from the plate to the cover(s) by radiation and convection, and from the top cover to the outside air by radiation and convection, but in different proportions. Heat losses through the insulated back and sides of the collector occur by conduction. Detailed methods of calculating all of these losses are described in Solar Energy Thermal Processes. Collectors should be designed so that all of these losses are as small as practical.

The product of the collector overall energy loss coefficient, U_L, and the temperature difference (T_i-T_a) occurring in Equation 2.1, represents the energy losses from the collector plate if it were all at the inlet fluid temperature. The collector plate will be at a higher temperature than the inlet fluid temperature when useful energy is being collected. This is necessary in order for heat to be transferred from the plate to the fluid. As a result, actual collector energy losses are higher than the product of U_L and (T_i-T_a). The difference is accounted for by the heat removal efficiency factor, F_R, which is described in greater detail in the next section.

The overall energy loss coefficient, U_L, is the sum of the loss coefficients corresponding to the top, bottom, and edge losses of the collector. For a well-designed collector, the sum of the bottom and edge loss coefficients is typically about 0.5 to 0.75 W/C per square meter of collector (2.8 to 4.2 BTU/hr-F per square foot of collector). The top loss coefficient is a function of the absorber plate temperature, the number of transparent covers, the cover material, the thermal (infrared) emittance of the absorber plate, the ambient temperature, and the windspeed. For most collector designs, U_L can be estimated from the graphs or the equations given in _Solar Energy Thermal Processes_. However, the best values of U_L are obtained from collector tests.

2.2-3 COLLECTOR EFFICIENCY FACTORS

Equation 2.1 is convenient for analysis of solar energy systems because the useful energy gain is found using the inlet fluid temperature. However, collector energy losses occur from the mean collector plate temperature which is always higher than the inlet fluid temperature when useful energy is being collected. The effect of the collector heat removal efficiency factor, F_R, is to reduce the calculated useful energy gain from what it would be if the whole collector were at the inlet fluid temperature to what it actually is with a fluid which increases in temperature as it flows through the collector.

F_R can be calculated for a wide range of collector geometries, as is shown in _Solar Energy Thermal Processes_. It is nearly independent of the solar radiation intensity and the collector plate and ambient temperatures, but it is a function of the fluid flowrate and the absorber plate design (thickness, material properties, tube spacing, etc.). The products, $F_R(\tau\alpha)_n$ and F_RU_L, which are the numbers needed in the following chapters, can readily be determined from the standard collector tests noted in the next section.

2.3 COLLECTOR TESTING AND DATA

Collectors are often tested by the procedure recommended by the National Bureau of Standards (Hill et al. (1976)). The procedure is to operate the collector on a test stand under steady conditions, i.e., the solar radiation, windspeed, and the ambient and inlet fluid temperatures are essentially constant for a period such that the fluid outlet temperature and the useful energy gain do not change appreciably with time. The conditions, including windspeed, should be representative of the conditions in which the collector will be used. Careful measurements are made of the incident radiation, the collector fluid flowrate, inlet and outlet fluid temperatures, and the ambient temperature.

The useful energy gain is given by

$$Q_u = A\,G\,C_p\,(T_i - T_o) \qquad 2.2$$

where

G is the collector fluid mass flowrate per unit collector area

C_p is the specific heat of the collector fluid

T_o is the outlet fluid temperature

The result of collector tests is usually given in terms of collector efficiency, η, defined as

$$\eta = Q_u/AI_T \qquad 2.3$$

where I_T is the measured radiation on the collector surface per unit area.

Collector tests are either performed outdoors on clear days near solar noon, or indoors using a solar simulator. In either case, the diffuse component of the radiation is small and radiation which strikes the collector surface is nearly at normal incidence. As a result, the transmittance-absorptance product resulting from the collector tests corresponds to beam radiation at normal incidence. This normal incidence transmittance-absorptance product is written as $(\tau\alpha)_n$.

The results of collector tests are best presented as a plot of instantaneous efficiency versus $(T_i-T_a)/I_T$. The theoretical basis for presenting the collector test results in this manner can be seen by dividing both sides of Equation 2.1 by I_TA. The instantaneous collector efficiency is then expressed

$$\eta = Q_u/AI_T = F_R(\tau\alpha)_n - F_RU_L(T_i-T_a)/I_T \qquad 2.4$$

If U_L is assumed to be constant, the plot of collector efficiency versus $(T_i-T_a)/I_T$ results in a straight line having a slope equal to $-F_RU_L$ and a vertical axis intercept equal to $F_R(\tau\alpha)_n$. This is most convenient since the values of F_RU_L and $F_R(\tau\alpha)_n$ are needed in Chapter 5 to estimate long-term solar heating system performance.

Collector test results are presented in this manner by Simon (1976) for many different flat-plate collectors. Typical collector test results plotted in this way for liquid collectors are shown in Figure 2.2. Generally the test data scatter about a straight line. The scatter is caused by variations of U_L with windspeed and temperature, as well as by errors in measurements. For the kinds of collectors normally used for solar heating, and for purposes of the f-chart design method, the test results can be adequately represented by a single straight line (implying that U_L can be considered to be relatively constant).

The major characteristics of the collectors which affect the slope and intercept of the efficiency versus $(T_i-T_a)/I_T$ curves are the number of covers and the nature of the absorber surface, i.e., whether it is a selective or a flat-black surface. The four typical collectors shown in Figure 2.2 are: A, 1 cover, flat-black surface; B, 2 covers, flat-black surface; C, 1 cover, selective surface; D, 2 covers, selective surface. Note that the relative efficiencies vary, depending on the temperature range of operation. It is not possible to say which will perform best in a system until the performance calculations described in Chapter 5 are completed. Note also that the cost of these collectors will vary; the methods of Chapter 6 must be used to determine which collector will result in the most economical system. (The curves in Figure 2.2 are shown to illustrate differences among collectors. Test data

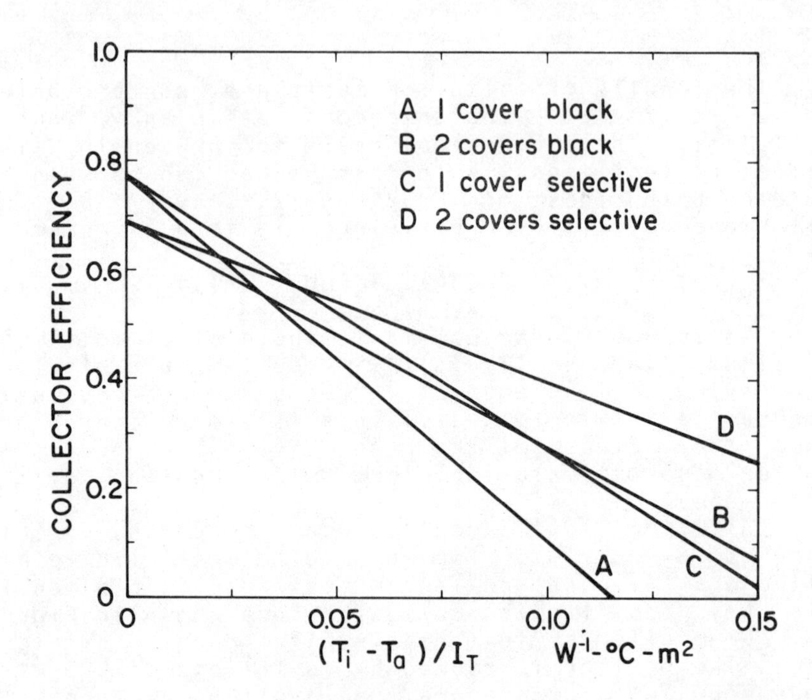

FIGURE 2.2
TEST RESULTS FOR FOUR LIQUID HEATING COLLECTORS

should be used to characterize a particular collector whenever possible.)

Collector efficiency is sometimes plotted against the difference between either the average fluid temperature or the outlet fluid temperature and the ambient temperature divided by I_T, that is, against $(T_{avg}-T_a)/I_T$ or $(T_o-T_a)/I_T$. Plots of this nature appear very similar to the plot of collector efficiency versus $(T_i-T_a)/I_T$; however, the slope and intercept of these plots have a different interpretation. If the mass flowrate of the fluid flowing through the collector during the test is known, then the values of F_RU_L and $F_R(\tau\alpha)_n$ can be determined from the slope and intercept of these test results. This is done by multiplying both the slope and the intercept of these plots by a factor K, where

$$K = \begin{cases} GC_p/(GC_p - \text{Slope}/2) & \text{for } \eta \text{ vs } (T_{avg}-T_a)/I_T \\ GC_p/(GC_p - \text{Slope}) & \text{for } \eta \text{ vs } (T_o-T_a)/I_T \end{cases} \quad 2.5$$

Then

$$\left.\begin{aligned} F_R U_L &= -K \text{ (Slope)} \\ F_R(\tau\alpha)_n &= K \text{ (Intercept)} \end{aligned}\right\} \quad 2.6$$

Note that the slope of the collector efficiency plot is a negative number and multiplication by -K results in a positive value for $F_R U_L$. The following examples demonstrate how $F_R U_L$ and $F_R(\tau\alpha)_n$ can be determined from collector test results.

EXAMPLE 2.1 Test Results of a Liquid Heating Collector

The performance curve for a liquid heating collector having two glass covers and a flat-black absorber surface is given in Figure 2.3. Determine the values of $F_R(\tau\alpha)_n$ and $F_R U_L$ for this collector.

The test data are presented as a plot of collector efficiency versus $(T_i - T_a)/I_T$. In this case, the value of $F_R(\tau\alpha)_n$ is the value of the Y-axis intercept of the plot, which is 0.68. The value of $F_R U_L$ is negative the value of the slope of the plot. Therefore, $F_R U_L$ is 3.75 W/C per square meter of collector.

Test results for air heaters are presented in an identical manner to that for liquid heating collectors. However, a distinction between liquid and air collector tests arises because F_R for liquid collectors remains relatively constant over a large range of collector flowrates, whereas there can be significant variation in F_R over the range of air flowrates used in air heating collectors. A different collector efficiency curve is obtained for each air flowrate. The designer of a solar heating system using air must be sure to use the $F_R(\tau\alpha)_n$ and $F_R U_L$ values corresponding to the air flowrate which will be used in the installed system. As with liquid collectors, test results are best presented in terms of

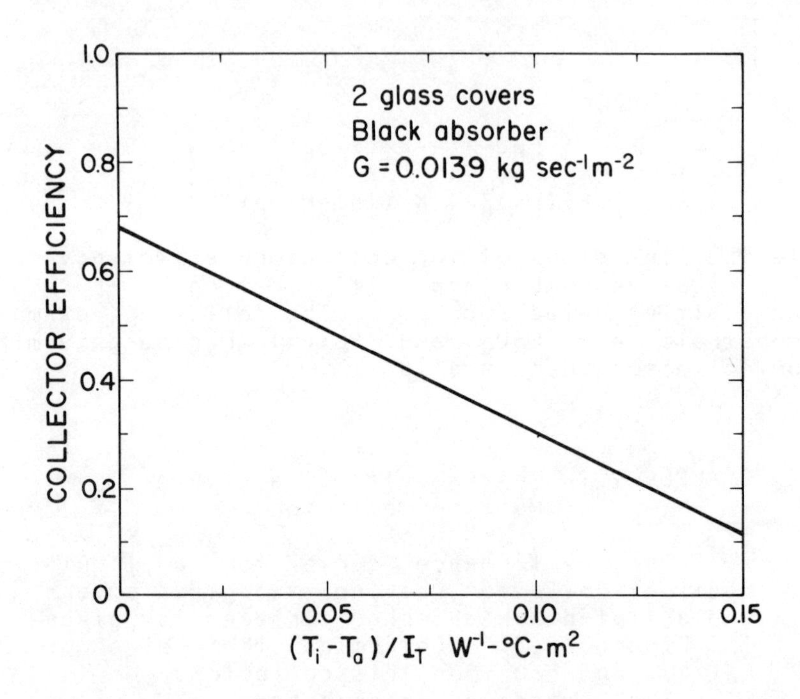

FIGURE 2.3
TEST PERFORMANCE OF A LIQUID HEATING COLLECTOR

$(T_i - T_a)/I_T$, but they are also commonly presented in other terms. In this case, the corrections required to obtain $F_R(\tau\alpha)_n$ and F_RU_L from the air heater tests are given by Equations 2.5 and 2.6.

EXAMPLE 2.2 Test Results of an Air Heater

Calculate the values of $F_R(\tau\alpha)_n$ and F_RU_L for an air heater having two glass covers and a flat-black absorber surface. The test results are given in Figure 2.4, plotted in terms of the outlet air temperature.

The experimental collector efficiency is given in terms of $(T_i - T_a)/I_T$. The values of F_RU_L and $F_R(\tau\alpha)_n$ are determined by multiplying the slope and intercept by K, where K is $GC_p/(GC_p - \text{Slope})$, as indicated in

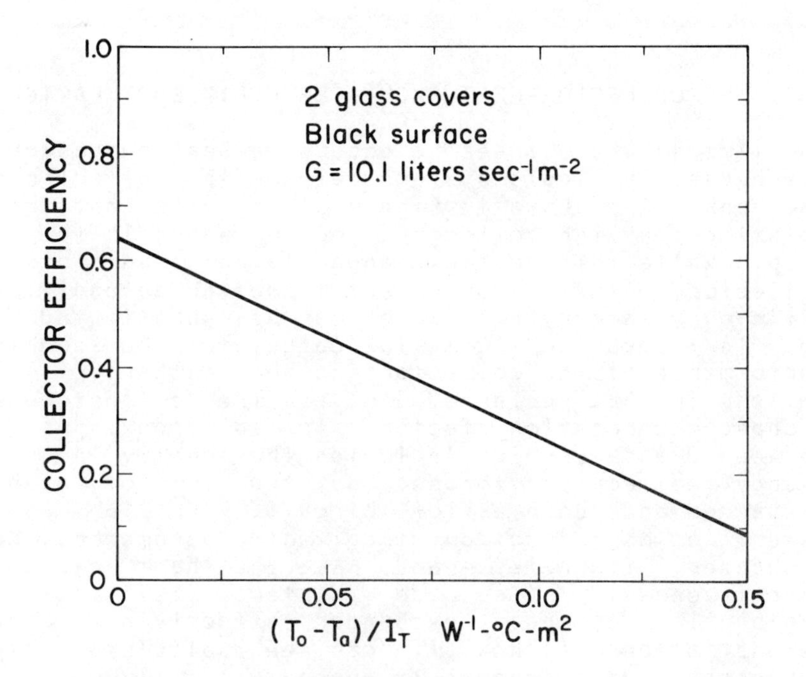

FIGURE 2.4
TEST PERFORMANCE OF A SOLAR AIR HEATER

Equation 2.5 for test data plotted in this manner. The collector mass flowrate, G, is 10.1 l/s per square meter of collector area. The specific heat of air is 1012 J/kg-C and the density of air at standard conditions is 0.001204 kg/liter. Thus GC_p is (10.1 l/s-m^2) x (1012 J/kg-C) x (0.001204 kg/l) = 12.3 W/C per square meter. The slope of the plot in Figure 2.4 is -3.7 W/C per square meter. Thus

$$K = GC_p/(GC_p - \text{Slope}) = 12.2/(12.2+3.7) = 0.77$$

From Equation 2.6

$$F_R(\tau\alpha)_n = K \times \text{Intercept} = 0.77 \times 0.64 = 0.49$$

$$F_R U_L = -K \times \text{Slope} = -0.77 \times (-3.7) = 2.85 \text{ W/C-m}^2$$

2.4 THE COLLECTOR-HEAT EXCHANGER EFFICIENCY FACTOR

In climates where freezing occurs, a heat exchanger is often used in liquid systems between the collector and the tank, as shown in Figure 1.1, with antifreeze solution in the collector loop and water in the tank loop. While this heat exchanger is not a part of the collector, it is convenient to define an additional efficiency factor, F_R', which can be substituted for F_R in Equation 2.1 to calculate the combined performance of the collector and heat exchanger. The ratio F_R'/F_R, referred to as the collector-heat exchanger correction factor, is an index, ranging between 0 and 1, which indicates the penalty in useful energy collection imposed by the use of a heat exchanger and double-flow circuit. F_R'/F_R can be determined as a function of collector parameters, heat exchanger flowrates, and ε_c, the heat exchanger effectiveness. (See de Winter (1975).) (An explanation of heat exchanger effectiveness and a demonstration of how it can be calculated from performance data appears in Appendix 1.)

$$F_R'/F_R = \left[1 + (F_R U_L/GC_p)(AGC_p/\varepsilon_c C_{min} - 1)\right]^{-1} \qquad 2.7$$

where C_{min} is the smaller of the two fluid capacitance rates (mass flowrate times fluid specific heat) in the heat exchanger. When the mass flowrates through the two sides of the heat exchanger are identical, C_{min} will be the capacitance rate of the fluid flowing through the collectors since the specific heat of the antifreeze solution in the collector is less than that of pure water. F_R'/F_R is plotted as a function of $F_R U_L/GC_p$ and $\varepsilon_c C_{min}/AGC_p$ in Figure 2.5.

EXAMPLE 2.3 Collector-Heat Exchanger Factor

The collectors described in Example 2.1 are to be used in a solar heating system that has a separate antifreeze, corrosion-inhibiting solution flowing in the collectors, as in Figure 1.1. Both flowrates in the collector-tank heat exchanger are to be 0.0139 kg/s per square meter of collector area. The specific heat is 3350 J/kg-C for

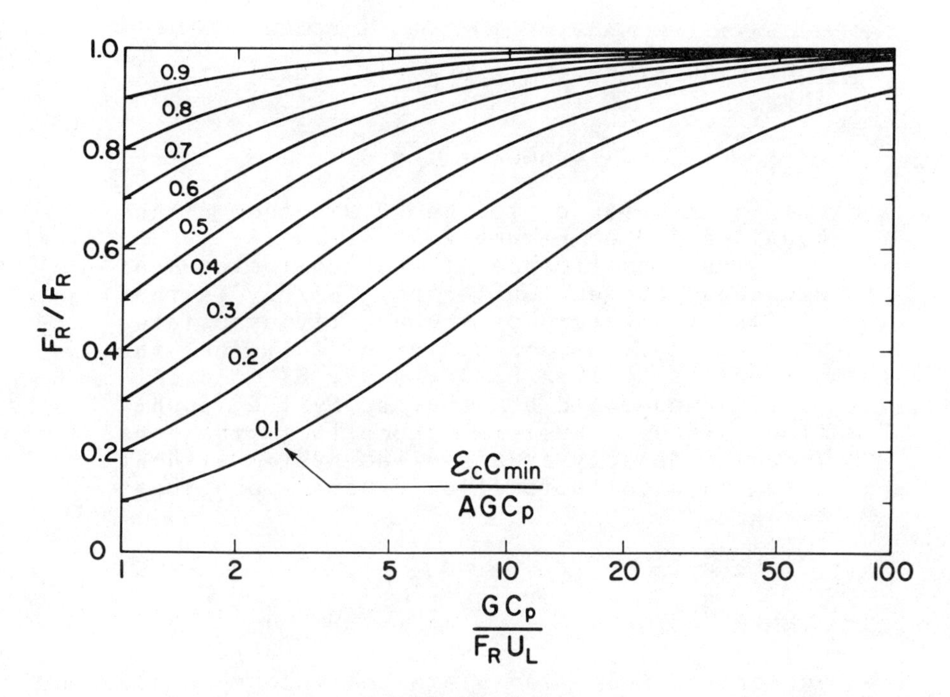

FIGURE 2.5
COLLECTOR-HEAT EXCHANGER CORRECTION FACTOR

the antifreeze solution and 4190 J/kg-C for water. The effectiveness of the heat exchanger will be assumed to be 0.7. Calculate the value of the collector-heat exchanger correction factor, F_R'/F_R.

From Example 2.1, F_RU_L = 3.75 W/C-m^2. The ratio of the capacitance rate in the collectors (per unit area) to F_RU_L is

$$GC_p/F_RU_L = (0.0139 \text{ kg/s-m}^2) \times (3350 \text{ J/kg-C}) / (3.75 \text{ W/C-m}^2) = 12.41$$

Since both flowrates in the heat exchanger are to be 0.0139 kg/s-m^2 and the specific heat of the antifreeze solution is less than that of water, the minimum capacitance rate in the heat exchanger, C_{min}, is that of the antifreeze solution flowing in the collector. Therefore, C_{min}/GC_p is equal to 1. Thus

$$\varepsilon_C C_{min}/AGC_p = \varepsilon_C = 0.7$$

F_R'/F_R is found to be 0.97 from either Equation 2.7 or Figure 2.5.

The significance of the collector-heat exchanger correction factor, F_R'/F_R, is that it is a measure of the penalty associated with having a separate flow circuit for the collectors. When F_R'/F_R=0.97, 3% more collector area would be required by the double flow circuit system to collect the same amount of energy as a similar system without a separate collector flow circuit and heat exchanger.

2.5 SUMMARY

The performance of flat-plate collectors is given by an energy balance equation, which relates the useful energy gain to the absorbed solar energy and the heat losses. Collector tests are used to measure the useful energy gain over a range of temperatures and radiation levels. The results of standard tests are best plotted as efficiency, η, against $(T_i-T_a)/I_T$. When this method of presenting test results is used, a straight line representation of the data for collectors of usual designs is adequate for purposes of the f-chart design method. The intercept of this line (where T_i=T_a) is $F_R(\tau\alpha)_n$ and the slope is $-F_RU_L$. If collector efficiency is plotted against $(T_{avg}-T_a)/I_T$ or $(T_o-T_a)/I_T$, the slope and intercept of these plots have different interpretations, but they can be used to determine $F_R(\tau\alpha)_n$ and F_RU_L if the collector fluid flowrate during the test is known.

If a heat exchanger is used between the collector and the tank in a liquid system, the collector-heat exchanger correction factor, F_R'/F_R, can be defined to account for the combined performance of the collectors and the heat exchanger. Thus, the performance of the collectors (and the heat exchanger, if present) can be characterized by two numbers, $F_R'(\tau\alpha)_n$ and $F_R'U_L$, which are needed in Chapter 5.

CHAPTER 3
EFFECT OF COLLECTOR ORIENTATION

3.1 INSTANTANEOUS VERSUS LONG-TERM PERFORMANCE

The collector equation (i.e., Equation 2.1) can be used to calculate instantaneous collector performance. However, the designer of a solar heating system is more concerned with long-term performance than with instantaneous performance. The collector equation can be used in computer simulations to determine long-term system performance by calculating the energy collection over short (e.g., hourly) periods and summing these over monthly or annual periods. However, a much simpler method has been developed for standard solar heating systems of the type shown in Figures 1.1 and 1.2.

From here on, we will be concerned only with the monthly performance of solar heating systems. In this chapter, we will discuss the effect of collector orientation on monthly solar heating performance. Collector orientation affects performance in two ways. Most importantly, it directly affects the amount of solar radiation incident on the collector surface. Presented in the next two sections is a method of estimating the monthly average radiation on surfaces of a wide range of orientations. The equations are presented in detail in Section 3.2 and tables developed from these equations are given in Section 3.3. In addition, collector orientation affects the transmittance of the transparent covers and the absorptance of the collector plate since both are functions of the angle at which radiation strikes the collector surface. A method of estimating the monthly average transmittance-absorptance product is presented in Sections 3.4 and 3.5.

3.2 CALCULATION OF RADIATION ON TILTED SURFACES

Monthly averages of the daily solar radiation incident on a horizontal surface are available for many locations. This information is tabulated for many North American locations in Appendix 2. However, radiation data on tilted surfaces are generally not available. A method of estimating the average daily radiation for each month on surfaces tilted directly towards the equator has been developed by Liu and Jordan (1962). This method has been compared with experimental data and extended to allow estimation of the monthly average daily radiation on surfaces oriented east or west of south by Klein (1976d).

The monthly average daily radiation on a tilted surface, $\bar{H}_T$, can be expressed

$$\bar{H}_T = \bar{R}\,\bar{H} \qquad 3.1$$

where

$\bar{H}$ is the monthly average daily radiation on a horizontal surface

$\bar{R}$ is the ratio of the monthly average daily radiation on a tilted surface to that on a horizontal surface for each month

$\bar{R}$ can be estimated by individually considering the beam, diffuse, and reflected components of the radiation. Assuming diffuse radiation to be isotropic (i.e., uniformily distributed over the sky), $\bar{R}$ can be expressed

$$\bar{R} = (1-\bar{H}_d/\bar{H})\bar{R}_b + \bar{H}_d/\bar{H}(1+\cos s)/2 + \rho(1-\cos s)/2 \qquad 3.2$$

where

$\bar{H}_d$ is the monthly average daily diffuse radiation

$\bar{R}_b$ is the ratio of the monthly average beam radiation on the tilted surface to that on a horizontal surface for each month

s is the tilt of the surface from horizontal

ρ is the ground reflectance. Liu and Jordan suggest that ρ varies from 0.2 to 0.7 depending upon the extent of snow cover.

In this equation, the first term is the contribution of the beam radiation, the second term is the contribution of diffuse radiation from the sky, and the third term is the contribution of radiation reflected into the collector from the ground.

Since measurements of $\overline{H}_d$, the monthly average daily diffuse radiation, are rarely available, $\overline{H}_d$ must be estimated from measurements of the average daily total radiation. Studies have shown that the fraction of the total radiation which is diffuse, $\overline{H}_d/\overline{H}$, is a function of $\overline{K}_T$, the ratio of actual daily radiation to the daily extraterrestrial radiation. This is shown in Figure 3.1. Extraterrestrial radiation is the solar radiation that would be received on a horizontal surface if there were no atmosphere. The daily extraterrestrial radiation can be calculated from a knowledge of the solar constant and geometric considerations. Monthly averages of the daily extraterrestrial radiation are shown in Table 3.1 for northern latitudes between 20° and 60°. The relationship in Figure 3.1 can be expressed analytically by

$$\overline{H}_d/\overline{H} = 1.39 - 4.03\ \overline{K}_T + 5.53\ \overline{K}_T^2 - 3.11\ \overline{K}_T^3 \qquad 3.3$$

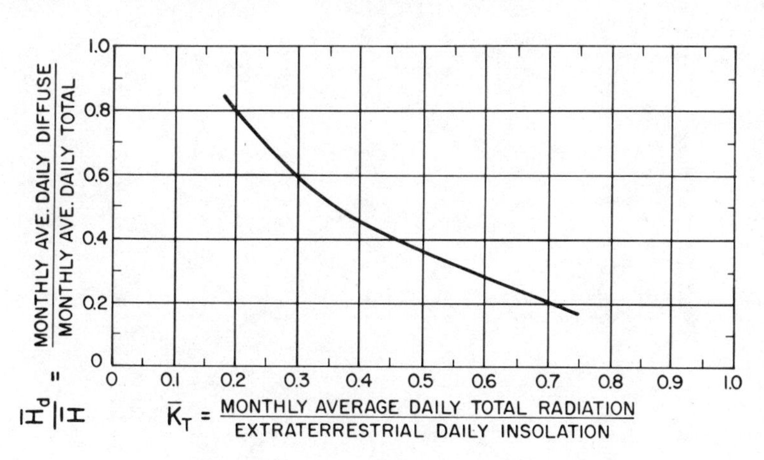

FIGURE 3.1
RELATIONSHIP OF THE DIFFUSE RADIATION FRACTION TO $\overline{K}_T$
(from Liu and Jordan (1960))

TABLE 3.1

MONTHLY AVERAGE DAILY EXTRATERRESTRIAL RADIATION MJ/m^2

LATITUDE	JAN	FEB	MAR	APR	MAY	JUN	JUL	AUG	SEP	OCT	NOV	DEC
25	23.9	28.2	33.0	37.1	39.4	40.1	39.6	37.9	34.4	29.5	24.9	22.7
30	21.1	25.7	31.3	36.5	39.6	40.7	40.1	37.6	33.1	27.3	22.1	19.7
35	18.1	23.1	29.3	35.5	39.6	41.2	40.3	37.0	31.5	24.9	19.2	16.7
40	15.1	20.3	27.2	34.3	39.3	41.4	40.3	36.2	29.7	22.3	16.3	13.6
45	12.0	17.5	24.8	32.8	38.8	41.3	40.0	35.1	27.7	19.6	13.3	10.6
50	9.0	14.5	22.3	31.2	38.1	41.2	39.6	33.8	25.4	16.7	10.3	7.6
55	6.1	11.5	19.5	29.3	37.2	40.9	39.1	32.4	23.0	13.8	7.3	4.8

In theory, $\bar{R}_b$, is a complicated function of the transmittance of the atmosphere. However, $\bar{R}_b$ can be estimated as the ratio of extraterrestrial radiation on the tilted surface to that on a horizontal surface for the month. For surfaces facing directly towards the equator, $\bar{R}_b$ is given as a function of ϕ, the latitude, and s, the collector slope, in Equation 3.4 and in Figures 3.2A through 3.2D. These values of $\bar{R}_b$ may be used for surfaces oriented as much as 15° east or west of south with little error. Values of $\bar{R}_b$ for surfaces oriented more than 15° away from south can be estimated by the method given by Klein (1976d).

$$\bar{R}_b = \frac{\cos(\phi - s)\cos\delta\sin\omega_s' + \pi/180\,\omega_s'\sin(\phi - s)\sin\delta}{\cos\phi\cos\delta\sin\omega_s + \pi/180\,\omega_s\sin\phi\sin\delta} \qquad 3.4$$

where

ω_s is the sunset hour angle on a horizontal surface given by

$$\omega_s = \arccos(-\tan\phi \times \tan\delta) \qquad 3.5$$

ω_s' is the sunset hour angle on the tilted surface given by

$$\omega_s' = \text{MIN}\,[\omega_s, \arccos(-\tan(\phi - s) \times \tan\delta)] \qquad 3.6$$

δ is the solar declination given by

$$\delta = 23.45\sin[360 \times (284+n)/365] \qquad 3.7$$

n is the day of the year

A worksheet has been developed to organize the calculations required to estimate $\bar{R}$ and the average daily radiation on tilted surfaces. The following example demonstrates the use of the worksheet. A blank worksheet is included in Appendix 5.

EXAMPLE 3.1 Radiation on a 58° Surface in Madison

Estimate the monthly averages of daily radiation incident on a south facing surface

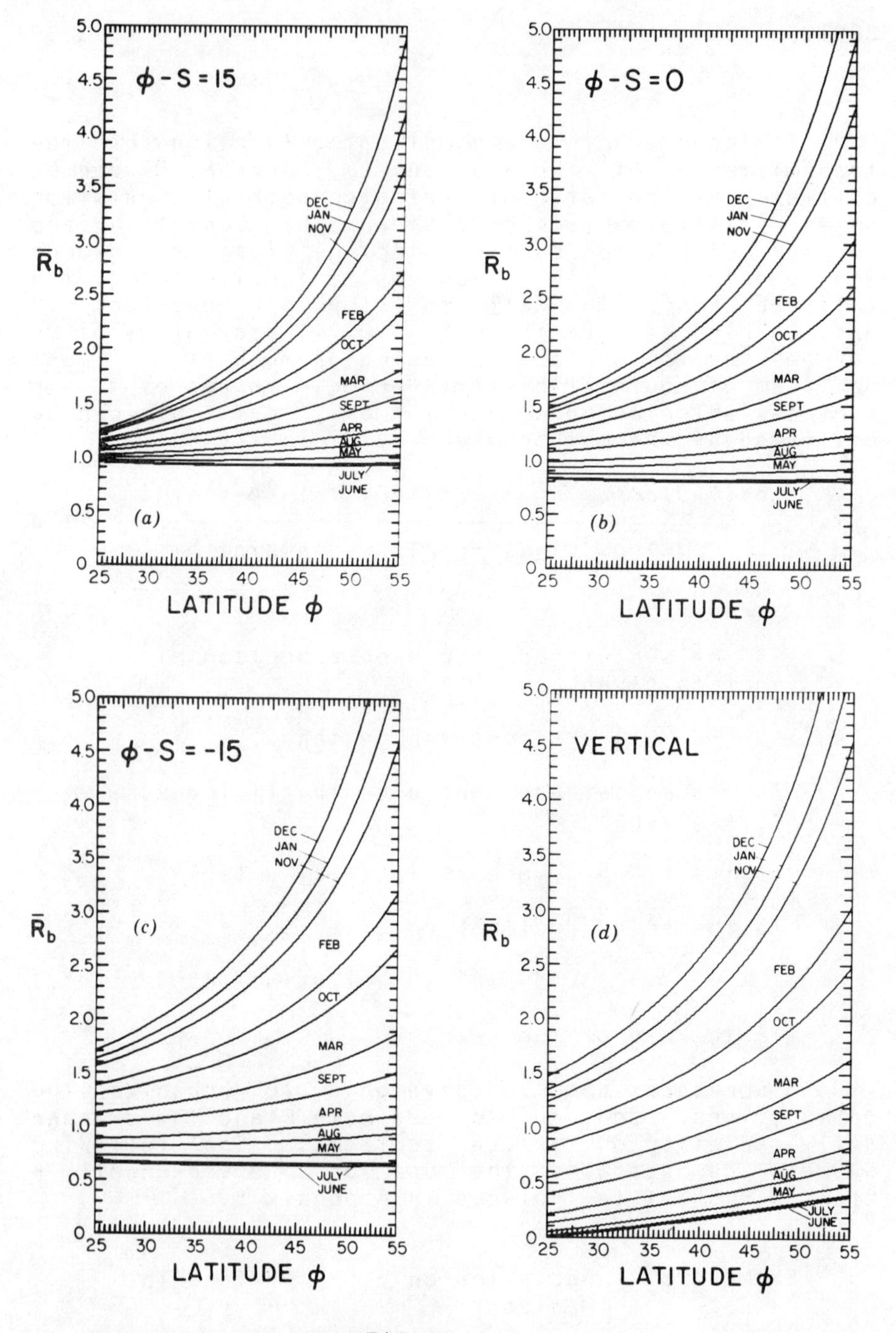

FIGURE 3.2
$\overline{R}_b$ FOR SOUTH FACING SURFACES

tilted 58° from horizontal in Madison, Wisconsin (lat. 43°N).

The calculations will follow the worksheet in Table 3.2. Daily average values of $\bar{H}$ (column G2) and corresponding values of $\bar{K}_T$ (column G3) for Madison can be found in Appendix 2. ($\bar{K}_T$ could also have been calculated by dividing $\bar{H}$ by the monthly average daily extraterrestrial radiation from Table 3.1).

The ratio of diffuse to total radiation, $\bar{H}_d/\bar{H}$ (column G4) is found for each month from Figure 3.1 or Equation 3.3. Values of $(1-\bar{H}_d/\bar{H})$ are tabulated in column G5.

$\bar{R}$ is the sum of three terms corresponding to the contributions of the beam, diffuse, and ground-reflected components of the radiation on the tilted surface. The contribution of the beam component is the product of $(1-\bar{H}_d/\bar{H})$ (column G5) and $\bar{R}_b$ (column G6), where $\bar{R}_b$ is found from Figure 3.2C or Equation 3.4. The contribution of the diffuse component (column G8) is the product of $\bar{H}_d/\bar{H}$ (column G4) and $(1+\cos s)/2$ (item D). $\bar{R}$ (column G9) is the sum of the contributions of the beam component (column G7), the diffuse component (column G8), and the reflected component (item F).

The average daily radiation on the tilted surface, $\bar{H}_T$ (column G10) is the product of $\bar{H}$ (column G2) and $\bar{R}$ (column G9).

3.3 $\bar{R}$ TABLES

$\bar{R}$ has been calculated from the equations of Section 3.2 for a range of values of latitude, slope, and $\bar{K}_T$. Tables 3.3A through 3.3E give $\bar{R}$ for a ground reflectance of 0.2. These tables will result in conservative estimates of $\bar{R}$ when there is snow cover and the ground reflectance is greater than 0.2. These tables greatly simplify the calculation of radiation on tilted surfaces, as demonstrated in the following example.

TABLE 3.2 WORKSHEET FOR EXAMPLE 3.1

COLLECTOR ORIENTATION WORKSHEET 1

AVERAGE DAILY RADIATION ON TILTED SURFACES

A. Location *Madison, Wisc.* B. Latitude ϕ = *43°* C. Inclination s = *58°*

D. (1+cos s)/2 *0.765* E. Ground Reflectance ρ = *0.2* F. ρ(1-cos s)/2 = *0.047*

G1. Month	G2. $\bar{H}$ J/Day-m^2 (Appendix 2)	G3. $\bar{K}_T$ (Appendix 2)	G4. $\bar{H}_d/\bar{H}$ Fig. 3.1 or Eqn. 3.3)	G5. $1-\bar{H}_d/\bar{H}$ (1-G4.)	G6. $\bar{R}_b$ (Fig. 3.2 or Eqn.3.4)	G7. Beam (G5.xG6.)	G8. Diffuse (D.xG4.)	G9. $\bar{R}$ (G7.+G8.+F.)	G10. $\bar{H}_T$ J/Day-m^2 (G9.xG2.)
Jan	6.41 x10^6	0.49	0.38	0.62	2.77	1.72	0.29	2.06	13.20 x10^6
Feb	9.22 x10^6	0.50	0.37	0.63	2.03	1.28	0.28	1.61	14.84 x10^6
Mar	13.99 x10^6	0.54	0.34	0.66	1.41	0.93	0.26	1.24	17.35 x10^6
Apr	16.53 x10^6	0.49	0.38	0.62	0.95	0.69	0.29	0.93	15.37 x10^6
May	19.82 x10^6	0.51	0.37	0.63	0.70	0.44	0.28	0.77	15.26 x10^6
Jun	23.07 x10^6	0.56	0.33	0.67	0.61	0.41	0.25	0.71	16.38 x10^6
Jul	23.24 x10^6	0.58	0.31	0.69	0.62	0.43	0.24	0.72	16.73 x10^6
Aug	19.76 x10^6	0.56	0.33	0.67	0.84	0.56	0.25	0.86	16.99 x10^6
Sep	16.40 x10^6	0.58	0.31	0.69	1.20	0.83	0.24	1.12	18.37 x10^6
Oct	11.28 x10^6	0.55	0.34	0.66	1.82	1.20	0.26	1.51	17.03 x10^6
Nov	6.31 x10^6	0.44	0.43	0.57	2.54	1.45	0.33	1.83	11.55 x10^6
Dec	5.63 x10^6	0.48	0.39	0.61	3.03	1.85	0.30	2.21	12.44 x10^6

TABLE 3.3A VALUES OF $\overline{R}$ FOR $\overline{K}_T = 0.30$

LATITUDE	JAN	FEB	MAR	APR	MAY	JUN	JUL	AUG	SEP	OCT	NOV	DEC
(LATITUDE-TILT)= 15.0												
25	1.09	1.06	1.03	1.00	.98	.98	.98	.99	1.02	1.05	1.08	1.09
30	1.15	1.10	1.05	1.01	.98	.97	.97	.99	1.03	1.08	1.13	1.16
35	1.23	1.15	1.07	1.01	.97	.96	.96	1.00	1.05	1.12	1.20	1.25
40	1.34	1.22	1.11	1.02	.97	.95	.96	1.00	1.07	1.18	1.30	1.38
45	1.51	1.31	1.15	1.03	.97	.94	.95	1.00	1.10	1.25	1.45	1.58
50	1.77	1.44	1.21	1.05	.97	.93	.95	1.01	1.14	1.35	1.67	1.91
55	2.24	1.65	1.29	1.07	.96	.93	.94	1.02	1.19	1.50	2.04	2.53
(LATITUDE-TILT)= .0												
25	1.17	1.11	1.04	.97	.93	.91	.92	.95	1.01	1.08	1.16	1.19
30	1.24	1.15	1.05	.97	.92	.90	.91	.95	1.02	1.11	1.21	1.27
35	1.33	1.20	1.08	.97	.91	.89	.90	.95	1.03	1.16	1.29	1.38
40	1.46	1.27	1.11	.98	.90	.87	.89	.94	1.05	1.21	1.41	1.53
45	1.65	1.37	1.15	.99	.90	.86	.88	.94	1.08	1.29	1.57	1.76
50	1.96	1.52	1.21	1.00	.89	.85	.87	.95	1.11	1.40	1.82	2.14
55	2.51	1.75	1.29	1.01	.89	.84	.86	.95	1.16	1.56	2.25	2.88
(LATITUDE-TILT)=-15.0												
25	1.21	1.11	1.00	.91	.84	.82	.83	.88	.96	1.07	1.18	1.24
30	1.28	1.15	1.01	.90	.83	.80	.81	.87	.97	1.10	1.24	1.32
35	1.37	1.20	1.03	.90	.82	.79	.80	.86	.97	1.14	1.32	1.43
40	1.51	1.27	1.06	.90	.81	.77	.79	.86	.99	1.19	1.44	1.60
45	1.71	1.37	1.10	.90	.80	.76	.77	.85	1.01	1.27	1.61	1.84
50	2.04	1.52	1.15	.91	.79	.74	.76	.85	1.04	1.38	1.88	2.26
55	2.63	1.76	1.23	.92	.78	.73	.75	.85	1.08	1.54	2.33	3.05
VERTICAL												
25	.94	.78	.62	.48	.42	.40	.41	.45	.56	.73	.90	.99
30	1.04	.85	.67	.52	.44	.42	.43	.48	.60	.79	.99	1.10
35	1.17	.94	.72	.55	.47	.44	.45	.51	.65	.86	1.10	1.24
40	1.33	1.04	.78	.59	.50	.47	.48	.55	.70	.95	1.25	1.44
45	1.57	1.18	.86	.64	.53	.49	.51	.59	.76	1.06	1.45	1.72
50	1.93	1.36	.95	.68	.56	.52	.54	.63	.82	1.20	1.75	2.17
55	2.55	1.62	1.06	.74	.60	.55	.57	.67	.91	1.40	2.24	3.00

TABLE 3.3B VALUES OF $\overline{R}$ FOR $\overline{K}_T$ = 0.40

LATITUDE	JAN	FEB	MAR	APR	MAY	JUN	JUL	AUG	SEP	OCT	NOV	DEC
(LATITUDE-TILT)= 15.0												
25	1.11	1.08	1.04	1.01	.98	.97	.98	1.00	1.03	1.07	1.10	1.13
30	1.20	1.13	1.07	1.01	.98	.96	.97	1.00	1.05	1.11	1.18	1.22
35	1.31	1.20	1.11	1.03	.97	.95	.96	1.00	1.07	1.17	1.28	1.34
40	1.46	1.30	1.15	1.04	.97	.94	.96	1.01	1.10	1.25	1.41	1.52
45	1.69	1.43	1.21	1.06	.97	.94	.95	1.02	1.15	1.35	1.61	1.79
50	2.04	1.61	1.30	1.09	.98	.94	.95	1.04	1.20	1.49	1.90	2.22
55	2.68	1.89	1.41	1.12	.98	.93	.95	1.06	1.28	1.70	2.41	3.06
(LATITUDE-TILT)= .0												
25	1.24	1.15	1.06	.98	.92	.90	.91	.95	1.03	1.12	1.22	1.27
30	1.34	1.21	1.09	.98	.91	.88	.90	.95	1.04	1.17	1.30	1.38
35	1.46	1.29	1.13	.99	.91	.87	.89	.95	1.07	1.23	1.41	1.52
40	1.64	1.39	1.17	1.00	.90	.86	.88	.96	1.10	1.31	1.57	1.73
45	1.90	1.53	1.23	1.02	.90	.86	.88	.96	1.14	1.42	1.79	2.04
50	2.32	1.74	1.32	1.04	.90	.85	.87	.98	1.19	1.58	2.13	2.56
55	3.05	2.04	1.43	1.07	.90	.84	.87	.99	1.27	1.80	2.71	3.54
(LATITUDE-TILT)=-15.0												
25	1.31	1.17	1.03	.91	.82	.79	.80	.87	.98	1.12	1.27	1.35
30	1.41	1.23	1.06	.91	.81	.77	.79	.86	.99	1.17	1.36	1.46
35	1.54	1.31	1.09	.91	.80	.76	.78	.86	1.01	1.23	1.47	1.62
40	1.73	1.41	1.13	.92	.80	.75	.77	.86	1.04	1.31	1.64	1.84
45	2.01	1.56	1.19	.93	.79	.74	.76	.87	1.08	1.42	1.87	2.18
50	2.45	1.77	1.27	.95	.79	.73	.76	.88	1.13	1.58	2.24	2.74
55	3.24	2.09	1.39	.98	.79	.72	.75	.89	1.19	1.81	2.85	3.80
VERTICAL												
25	1.05	.84	.63	.44	.36	.34	.35	.40	.54	.77	.99	1.12
30	1.18	.94	.69	.49	.39	.36	.37	.44	.60	.85	1.11	1.26
35	1.35	1.05	.76	.54	.43	.39	.41	.49	.66	.95	1.26	1.45
40	1.57	1.18	.84	.59	.47	.42	.44	.53	.73	1.06	1.46	1.71
45	1.88	1.36	.94	.65	.51	.46	.48	.58	.81	1.21	1.73	2.08
50	2.36	1.60	1.06	.71	.55	.50	.52	.63	.90	1.40	2.12	2.68
55	3.18	1.95	1.21	.78	.60	.54	.56	.69	1.00	1.66	2.76	3.78

TABLE 3.3C VALUES OF $\overline{R}$ FOR $\overline{K}_T$ = 0.50

LATITUDE	JAN	FEB	MAR	APR	MAY	JUN	JUL	AUG	SEP	OCT	NOV	DEC
(LATITUDE-TILT)= 15.0												
25	1.14	1.09	1.05	1.01	.98	.97	.97	1.00	1.03	1.08	1.12	1.15
30	1.23	1.16	1.08	1.02	.97	.96	.96	1.00	1.06	1.13	1.21	1.26
35	1.37	1.24	1.13	1.03	.97	.95	.96	1.01	1.09	1.20	1.33	1.41
40	1.55	1.36	1.19	1.05	.97	.94	.96	1.02	1.13	1.30	1.49	1.62
45	1.82	1.51	1.26	1.08	.98	.94	.96	1.03	1.18	1.42	1.72	1.93
50	2.24	1.73	1.36	1.12	.99	.94	.96	1.06	1.25	1.59	2.08	2.45
55	2.99	2.06	1.50	1.16	1.00	.94	.96	1.08	1.34	1.83	2.67	3.44
(LATITUDE-TILT)= .0												
25	1.29	1.19	1.08	.98	.91	.88	.90	.95	1.04	1.15	1.26	1.32
30	1.40	1.26	1.11	.99	.91	.87	.89	.95	1.06	1.21	1.36	1.45
35	1.56	1.35	1.16	1.00	.90	.86	.88	.96	1.09	1.28	1.50	1.63
40	1.77	1.48	1.22	1.02	.90	.86	.88	.97	1.13	1.38	1.68	1.87
45	2.08	1.65	1.30	1.04	.90	.85	.87	.98	1.18	1.52	1.95	2.25
50	2.57	1.89	1.40	1.08	.91	.85	.87	1.00	1.25	1.70	2.36	2.86
55	3.44	2.26	1.54	1.12	.92	.85	.88	1.02	1.34	1.97	3.04	4.02
(LATITUDE-TILT)=-15.0												
25	1.38	1.22	1.05	.91	.81	.77	.79	.86	.99	1.16	1.33	1.43
30	1.50	1.29	1.09	.91	.80	.76	.78	.86	1.01	1.22	1.44	1.57
35	1.66	1.39	1.13	.92	.80	.75	.77	.86	1.04	1.30	1.58	1.75
40	1.89	1.52	1.19	.94	.79	.74	.76	.87	1.08	1.40	1.78	2.02
45	2.22	1.69	1.26	.96	.79	.73	.76	.88	1.12	1.53	2.06	2.43
50	2.75	1.94	1.36	.98	.79	.73	.76	.89	1.19	1.72	2.49	3.09
55	3.68	2.32	1.50	1.02	.80	.72	.75	.91	1.27	1.99	3.22	4.34
VERTICAL												
25	1.13	.89	.63	.42	.32	.29	.30	.37	.53	.80	1.06	1.21
30	1.29	1.00	.71	.47	.35	.32	.33	.41	.60	.89	1.20	1.38
35	1.48	1.13	.79	.53	.40	.35	.37	.47	.67	1.01	1.38	1.60
40	1.74	1.29	.89	.59	.44	.39	.41	.52	.75	1.14	1.61	1.91
45	2.11	1.50	1.00	.66	.49	.44	.46	.58	.84	1.31	1.92	2.34
50	2.67	1.78	1.14	.73	.54	.48	.51	.64	.95	1.54	2.39	3.04
55	3.64	2.19	1.32	.81	.60	.53	.56	.71	1.08	1.84	3.15	4.34

TABLE 3.3D VALUES OF $\overline{R}$ FOR $\overline{K}_T$ = 0.60

LATITUDE	JAN	FEB	MAR	APR	MAY	JUN	JUL	AUG	SEP	OCT	NOV	DEC
					(LATITUDE-TILT)= 15.0							
25	1.15	1.11	1.06	1.01	.98	.96	.97	1.00	1.04	1.09	1.14	1.17
30	1.27	1.18	1.10	1.02	.97	.95	.96	1.00	1.07	1.15	1.24	1.29
35	1.41	1.28	1.15	1.04	.97	.94	.96	1.01	1.10	1.23	1.38	1.46
40	1.62	1.41	1.21	1.07	.98	.94	.95	1.02	1.15	1.34	1.56	1.70
45	1.92	1.58	1.30	1.10	.98	.94	.96	1.04	1.21	1.48	1.82	2.05
50	2.40	1.83	1.41	1.14	.99	.94	.96	1.07	1.29	1.67	2.22	2.64
55	3.24	2.20	1.57	1.19	1.01	.94	.97	1.10	1.39	1.95	2.89	3.75
					(LATITUDE-TILT)= .0							
25	1.33	1.21	1.09	.98	.91	.87	.89	.95	1.05	1.17	1.30	1.37
30	1.46	1.30	1.13	.99	.90	.86	.88	.95	1.08	1.24	1.42	1.51
35	1.63	1.40	1.19	1.01	.90	.85	.87	.96	1.11	1.33	1.57	1.71
40	1.88	1.55	1.26	1.03	.90	.85	.87	.97	1.16	1.44	1.78	1.99
45	2.23	1.74	1.35	1.06	.91	.85	.87	.99	1.22	1.59	2.08	2.41
50	2.78	2.02	1.47	1.10	.92	.85	.88	1.02	1.30	1.81	2.54	3.10
55	3.76	2.43	1.63	1.15	.93	.85	.88	1.05	1.41	2.11	3.31	4.41
					(LATITUDE-TILT)=-15.0							
25	1.43	1.26	1.07	.91	.80	.75	.77	.86	1.00	1.19	1.39	1.49
30	1.57	1.34	1.11	.92	.79	.74	.76	.86	1.03	1.26	1.51	1.65
35	1.76	1.45	1.16	.93	.79	.73	.76	.86	1.06	1.35	1.67	1.86
40	2.02	1.60	1.23	.95	.79	.73	.75	.87	1.11	1.47	1.90	2.17
45	2.40	1.80	1.32	.98	.79	.72	.75	.89	1.16	1.62	2.22	2.63
50	2.99	2.09	1.44	1.01	.80	.72	.75	.91	1.24	1.84	2.71	3.37
55	4.04	2.52	1.59	1.05	.81	.72	.76	.93	1.34	2.15	3.52	4.78
					VERTICAL							
25	1.20	.92	.63	.39	.28	.25	.26	.34	.53	.82	1.12	1.28
30	1.37	1.04	.72	.46	.32	.28	.30	.39	.60	.93	1.28	1.48
35	1.59	1.19	.81	.52	.37	.32	.34	.45	.68	1.06	1.48	1.73
40	1.88	1.37	.92	.59	.42	.37	.39	.51	.77	1.21	1.73	2.07
45	2.30	1.61	1.05	.66	.48	.42	.44	.58	.87	1.40	2.09	2.56
50	2.93	1.93	1.21	.75	.54	.47	.50	.65	.99	1.65	2.61	3.34
55	4.01	2.39	1.41	.84	.60	.52	.55	.72	1.13	2.00	3.46	4.80

TABLE 3.3E VALUES OF $\overline{R}$ FOR $\overline{K}_T = 0.70$

LATITUDE	JAN	FEB	MAR	APR	MAY	JUN	JUL	AUG	SEP	OCT	NOV	DEC
					(LATITUDE-TILT)= 15.0							
25	1.17	1.12	1.06	1.01	.98	.96	.97	1.00	1.04	1.10	1.16	1.19
30	1.30	1.20	1.11	1.03	.97	.95	.96	1.00	1.07	1.17	1.27	1.33
35	1.46	1.31	1.17	1.05	.97	.94	.95	1.01	1.12	1.26	1.42	1.51
40	1.69	1.45	1.24	1.08	.98	.94	.95	1.03	1.17	1.38	1.62	1.78
45	2.03	1.65	1.34	1.12	.99	.94	.96	1.06	1.24	1.53	1.92	2.18
50	2.56	1.93	1.47	1.16	1.00	.94	.97	1.09	1.33	1.75	2.36	2.83
55	3.50	2.34	1.64	1.22	1.02	.94	.98	1.13	1.45	2.06	3.11	4.07
					(LATITUDE-TILT)= .0							
25	1.37	1.24	1.11	.98	.90	.86	.88	.95	1.06	1.20	1.34	1.41
30	1.52	1.34	1.16	1.00	.90	.85	.87	.95	1.09	1.27	1.47	1.58
35	1.71	1.46	1.22	1.02	.90	.85	.87	.96	1.13	1.37	1.64	1.80
40	1.98	1.62	1.30	1.05	.90	.84	.87	.98	1.19	1.50	1.88	2.11
45	2.38	1.84	1.40	1.08	.91	.84	.87	1.00	1.26	1.67	2.21	2.58
50	3.00	2.15	1.53	1.13	.92	.85	.88	1.03	1.35	1.91	2.73	3.35
55	4.09	2.61	1.72	1.19	.94	.85	.89	1.07	1.47	2.25	3.59	4.81
					(LATITUDE-TILT)=-15.0							
25	1.49	1.30	1.09	.91	.78	.73	.76	.85	1.01	1.23	1.44	1.56
30	1.65	1.39	1.14	.92	.78	.72	.75	.86	1.04	1.30	1.58	1.73
35	1.86	1.52	1.20	.94	.78	.72	.75	.87	1.09	1.41	1.76	1.98
40	2.15	1.69	1.28	.96	.78	.72	.74	.88	1.14	1.54	2.02	2.32
45	2.58	1.91	1.38	.99	.79	.71	.75	.90	1.20	1.71	2.37	2.83
50	3.24	2.23	1.51	1.04	.80	.72	.75	.92	1.29	1.96	2.92	3.66
55	4.41	2.72	1.69	1.09	.81	.72	.76	.96	1.40	2.31	3.83	5.23
						VERTICAL						
25	1.26	.96	.64	.37	.25	.21	.23	.31	.52	.85	1.18	1.36
30	1.46	1.09	.73	.44	.29	.25	.27	.37	.60	.97	1.35	1.57
35	1.70	1.26	.84	.51	.35	.29	.32	.43	.69	1.11	1.57	1.85
40	2.03	1.46	.96	.59	.40	.34	.37	.50	.79	1.28	1.86	2.23
45	2.49	1.72	1.10	.67	.47	.40	.43	.57	.90	1.49	2.25	2.78
50	3.19	2.07	1.28	.76	.53	.45	.48	.65	1.03	1.77	2.83	3.65
55	4.39	2.59	1.50	.87	.60	.51	.55	.74	1.19	2.15	3.78	5.27

EXAMPLE 3.2 Use of the $\bar{R}$ Tables

Estimate the monthly average daily radiation incident on a south facing surface tilted 58° from horizontal in Madison (lat. 43°N) using the $\bar{R}$ tables.

Values of $\bar{R}$ can be found from Tables 3.3A through E using linear interpolation where required. For example, $\bar{K}_T$ for January in Madison is 0.49. In Table 3.3B ($\bar{K}_T$=0.4) for January with (ϕ-s)=-15, $\bar{R}$ is 1.73 at a 40° latitude and 2.01 at a 45° latitude. In Table 3.3C ($\bar{K}_T$=0.5), $\bar{R}$ is 1.89 at a 40° latitude and 2.22 at a 45° latitude. Interpolating for $\bar{K}_T$=0.49, $\bar{R}$ is 1.87 at a 40° latitude and 2.20 at a 45° latitude. A second interpolation of these values for a latitude of 43° results in $\bar{R}$=2.07, which agrees with the value calculated in Example 3.1 using the detailed equations. Values of $\bar{R}$ for other months can be estimated in the same manner.

3.4 EFFECT OF ORIENTATION ON TRANSMITTANCE AND ABSORPTANCE - SHORTCUT METHOD

Both τ, the transmittance of the transparent collector cover system, and α, the absorptance of the collector plate, depend on the angle at which solar radiation strikes the collector surface. Separate values for τ and α cannot be determined by the collector test procedure described in Section 2.3; only the product of F_R, τ, and α is determined from the tests. This, however, is not a problem since it is the product of these three terms which is needed to estimate long-term solar heating system performance.

A problem that does arise is that collector tests are usually carried out with the radiation incident on the collector in a neary perpendicular direction. Thus, the product of F_R, τ, and α determined from collector tests ordinarily corresponds to the transmittance and absorptance values for radiation at normal incidence, $F_R(\tau\alpha)_n$. Depending on the collector orientation and the time of the year, the monthly average values of the transmittance and absorptance can be significantly lower than the values for radiation at normal incidence.

A shortcut method of determining the monthly average transmittance-absorptance product, $(\overline{\tau\alpha})$, which is useful for most common situations, is as follows. When a collector is oriented with a slope equal to the latitude plus or minus 15°, and when the collector faces within 15° of due south, the ratio, $(\overline{\tau\alpha})/(\tau\alpha)_n$, is about 0.96 for a single cover collector and 0.94 for a two-cover collector for all months during the heating season.

3.5 EFFECT OF ORIENTATION ON TRANSMITTANCE AND ABSORPTANCE - DETAILED METHOD

For a determination of $(\overline{\tau\alpha})/(\tau\alpha)_n$ for collectors tilted more than 15° from the latitude, the following analysis must be used. Radiation incident on the collector consists of beam, diffuse, and ground reflected components. As a result, the ratio of the monthly average transmittance-absorptance product, $(\overline{\tau\alpha})$, to the transmittance-absorptance product at normal incidence, $(\tau\alpha)_n$, can be calculated as a function of a weighted average for the beam, diffuse, and reflected components of the radiation. In a manner analogous to Equation 3.3,

$$(\overline{\tau\alpha})/(\tau\alpha)_n = (1-\overline{H}_d/\overline{H}) \times \overline{R}_b/\overline{R} \times (\overline{\tau\alpha})_b/(\tau\alpha)_n$$

$$+ \overline{H}_d/\overline{H} \times 1/\overline{R} \times (1+\cos s)/2 \times (\overline{\tau\alpha})_d/(\tau\alpha)_n$$

$$+ \rho \times 1/\overline{R} \times (1-\cos s)/2 \times (\overline{\tau\alpha})_r/(\tau\alpha)_n \qquad 3.8$$

where $(\overline{\tau\alpha})_b$, $(\overline{\tau\alpha})_d$, and $(\overline{\tau\alpha})_r$ are the monthly average values of the transmittance-absorptance product corresponding to beam, diffuse, and ground-reflected radiation.

In Figure 3.3, the ratio of the transmittance of the collector cover system for radiation at a known incidence angle to the transmittance for radiation at normal incidence is given for 1, 2, and 3 sheets of glass or Tedlar. In Figure 3.4, the ratio of the collector plate absorptance for solar radiation at a known incidence angle to that for radiation at normal incidence is given for a flat-black surface. The limited data available suggests that the angular dependence of selective surfaces is similar to that of the flat-black surface (Pettit and Sowell (1976)). With this information, it is necessary only to specify the monthly average incidence angles for beam, diffuse, and reflected radiation in order to determine $(\overline{\tau\alpha})/(\tau\alpha)_n$.

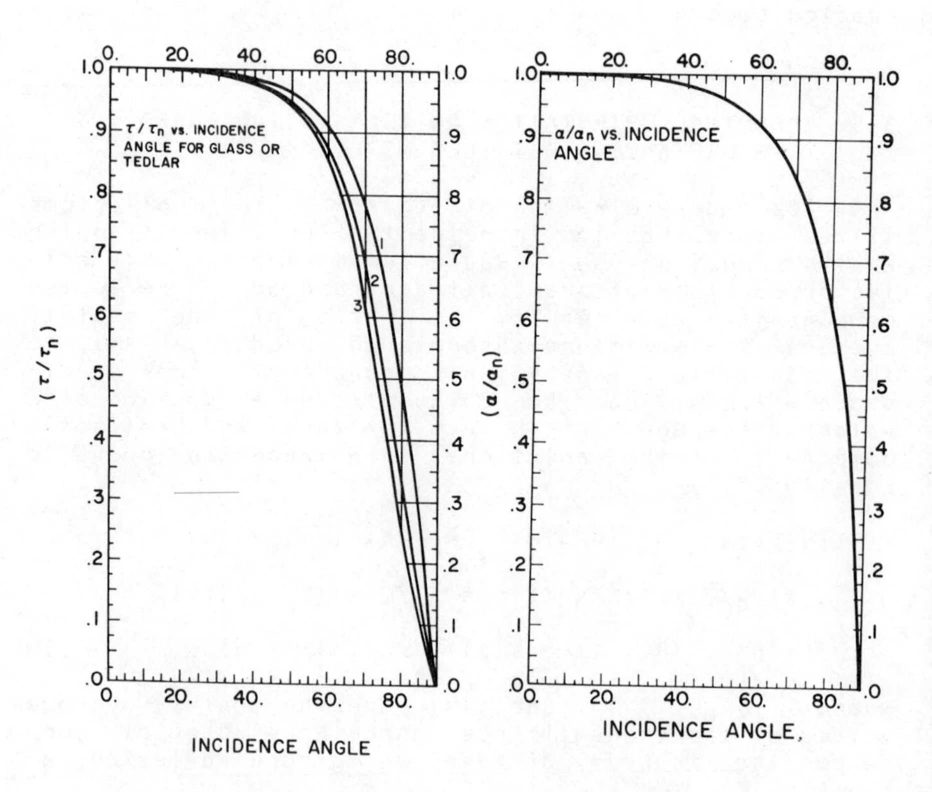

FIGURE 3.3
τ/τ_n VS INCIDENCE ANGLE

FIGURE 3.4
α/α_n VS INCIDENCE ANGLE

For isotropic diffuse radiation on a horizontal surface, the average angle of incidence is 60°. If the collector is tilted, the average angle of incidence for diffuse radiation will be less the 60°, while the average angle of incidence for the ground-reflected radiation will be more than 60°. As a conservative assumption, the average angle for diffuse radiation may be taken as 60°. The small contribution of ground-reflected radiation is also taken as having an average incidence angle of 60°. Then $(\overline{\tau\alpha})_d/(\tau\alpha)_n$ and $(\overline{\tau\alpha})_r/(\tau\alpha)_n$ can be evaluated in the same manner.

For surfaces facing directly towards the equator, Klein (1976a) has found that $\overline{\theta}_b$, the average incidence angle for beam radiation is approximately the angle at which radiation strikes the collector surface 2.5 hours from solar noon on a day in the middle of the month. $\overline{\theta}_b$ is given in Figure 3.5 as a function of $\phi - s$, the difference between the latitude and the collector slope. $(\overline{\tau\alpha})_b/(\tau\alpha)_n$ is the product of τ/τ_n and α/α_n corresponding to an incidence angle of $\overline{\theta}_b$.

No simple method for determining the value of $(\overline{\tau\alpha})/(\tau\alpha)_n$ has been found for surfaces facing more than 15° east or west of south. However, the value of $(\overline{\tau\alpha})/(\tau\alpha)_n$ is not a strong function of this azimuth angle. The value of $(\overline{\tau\alpha})/(\tau\alpha)_n$ calculated for surfaces facing due south can be used for surfaces with an azimuth angle as much as 15° with little error.

The procedure is then to calculate $(\overline{\tau\alpha})/(\tau\alpha)_n$ from Equation 3.8 using the values of $\overline{R}_b$, $\overline{R}$, and $\overline{H}_d/\overline{H}$ calculated in the manner described in the preceding section. The product $F_R(\overline{\tau\alpha})$, needed to estimate long-term system performance, is determined by multiplying the value of $F_R(\tau\alpha)_n$ obtained from the collector tests by $(\overline{\tau\alpha})/(\tau\alpha)_n$. A worksheet has been provided to organize the calculation of $(\overline{\tau\alpha})/(\tau\alpha)_n$. The use of this worksheet is illustrated in the following example. A blank worksheet is included in Appendix 5.

EXAMPLE 3.3 $(\overline{\tau\alpha})/(\tau\alpha)_n$ for a Two-Cover Collector at 58° in Madison

Flat-plate solar collectors having two glass covers and a flat-black absorber surface are being considered for a solar heating system in Madison, Wisconsin (lat. 43°N). The roof that they will be mounted on will be inclined at a 58° angle.

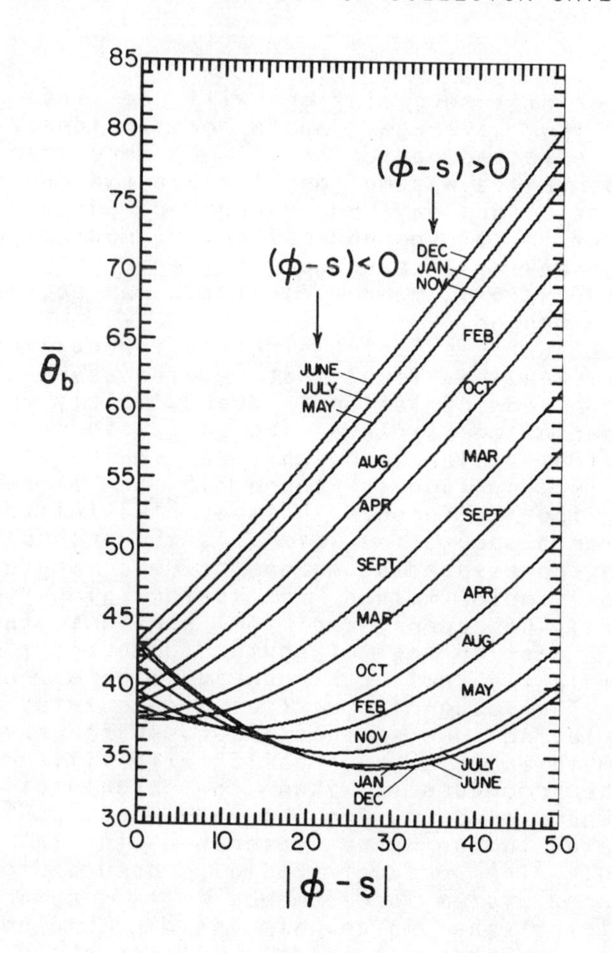

FIGURE 3.5
MEAN INCIDENCE ANGLE VS LATITUDE-COLLECTOR TILT

Calculate the monthly values of $(\overline{\tau\alpha})/(\tau\alpha)_n$ for this collector.

The worksheet in Table 3.4 will be used for the following calculations. Also, some of the results from Table 3.2, the worksheet used to calculate radiation on the 58° surface, will be needed here.

The monthly incidence angle for beam radiation, $\overline{\theta}_b$, is found from Figure 3.5 as a function of the difference between the latitude and the collector slope, which is

(43°-58°), or -15°. $\overline{\theta}_b$ is tabulated in column H2. τ/τ_n and α/α_n for beam radiation (columns H3 and H4) are determined from Figures 3.3 and 3.4 corresponding to an incidence angle of $\overline{\theta}_b$. The first term in Equation 3.4, for beam radiation, tabulated in column H6, is the product of $(1-\overline{H}_d/\overline{H})$ (column G5 from Table 3.2), $\overline{R}_b/\overline{R}$ (column G6 divided by column G9), τ/τ_n (column H3), and α/α_n (column H4).

Diffuse radiation is assumed to strike the collector at a mean angle of 60°. τ/τ_n for diffuse radiation is found from Figure 3.3 and tabulated in column H7. α/α_n is approximately 0.92 at 60°. The second term in Equation 3.4 (for diffuse radiation), tabulated in column H8, is the product of (1+cos s)/2 (item D from Table 3.2), $\overline{H}_d/\overline{H}$ (column G4), τ/τ_n (column H7), and α/α_n (0.92) divided by $\overline{R}$ (column G9).

The third term, for reflected radiation, column H9, is the product of ρ(1-cos s)/2 (item F from Table 3.2), τ/τ_n (column H7), α/α_n (0.92) divided by $\overline{R}$ (column G9).

$(\overline{\tau\alpha})/(\tau\alpha)_n$ (column H10) is the sum of the beam, diffuse and reflected terms in columns H6, H8, and H9. Note that $(\overline{\tau\alpha})/(\tau\alpha)_n$ is approximately 0.94 for a two-cover collector for most of the heating season when the collector is tilted within 15° of the latitude.

3.6 OPTIMUM COLLECTOR ORIENTATION

Using the calculation procedures described in Chapter 5, it is possible to evaluate the long-term thermal performance of solar heating systems for a range of collector orientations in order to find the optimum orientation, that is, the orientation in which the solar heating system provides the largest fraction of the annual heating load. This does not necessarily correspond to a situation in which the annual solar radiation on the collector surface is maximized, since

TABLE 3.4 WORKSHEET FOR EXAMPLE 3.3

COLLECTOR ORIENTATION WORKSHEET 2

MONTHLY AVERAGE TRANSMITTANCE - ABSORPTANCE PRODUCT

H1.	H2.	H3.	H4.	H5.	H6.	H7.	H8.	H9.	H10.
Month	$\overline{\theta}_b$ (Fig. 3.5)	τ/τ_n @ θ_b (Fig. 3.3)	α/α_n @ θ_b (Fig. 3.4)	$\overline{R}_b/\overline{R}$ (G6./G9.)	Beam (G5.xH3.x H4.xH5.)	τ/τ_n @ 60° (Fig. 3.3)	Diffuse (D.xG4./G9. x0.92xH7.)	Reflected (F./G9.x0.92 xH7.)	$(\overline{\tau\alpha})/(\tau\alpha)_n$ (H6.+H8.+H9.)
Jan	36	0.98	0.99	1.34	0.81	0.87	0.11	0.02	0.94
Feb	36	0.98	0.99	1.26	0.78	0.87	0.14	0.02	0.94
Mar	39	0.98	0.99	1.14	0.73	0.87	0.17	0.03	0.93
Apr	45	0.97	0.98	1.02	0.62	0.87	0.25	0.04	0.91
May	50	0.97	0.98	0.91	0.56	0.87	0.29	0.05	0.90
Jun	53	0.97	0.98	0.86	0.55	0.87	0.29	0.05	0.89
Jul	51	0.97	0.98	0.86	0.55	0.87	0.27	0.05	0.90
Aug	46	0.97	0.98	0.98	0.63	0.87	0.23	0.04	0.90
Sep	41	0.98	0.98	1.07	0.72	0.87	0.17	0.03	0.92
Oct	37	0.98	0.99	1.21	0.77	0.87	0.14	0.02	0.93
Nov	36	0.98	0.99	1.39	0.78	0.87	0.14	0.02	0.94
Dec	36	0.98	0.99	1.37	0.81	0.87	0.11	0.02	0.94

the relative time distribution of the solar radiation and the heating loads is an important factor. (See Figure 3.6.) For solar space heating systems, the optimum collector orientation is directly towards the equator (south in the northern hemisphere) at an angle 10° to 15° degrees greater than the latitude. For domestic water heating systems in which the heating load is approximately the same magnitude for all months, the optimum angle is about equal to the latitude. However, the collector orientation is not critical. Deviations from the optimum by as much as 15° have little effect on the annual performance of solar heating systems, as shown for a particular case in Figure 3.6.

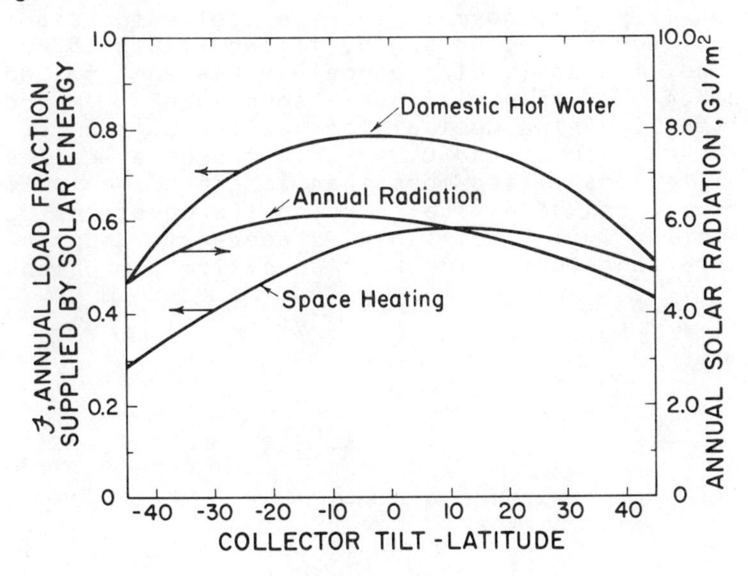

FIGURE 3.6
EFFECT OF COLLECTOR ORIENTATION

3.7 SUMMARY

Records of the long-term average daily radiation on a horizontal surface are given in Appendix 2 for many

North American locations. The solar radiation on a tilted surface is determined by multiplying the radiation on the horizontal surface by $\bar{R}$, where, $\bar{R}$ is the ratio of average daily radiation on the tilted surface to that on the horizontal surface. $\bar{R}$ can be determined using either the detailed equations presented in Section 3.2, or the $\bar{R}$ tables in Section 3.3. For hand calculations, the $\bar{R}$ tables will usually be more convenient unless the equations in Section 3.2 are evaluated using a programmable calculator.

Collector orientation also affects the transmittance of the collector cover(s) and the absorptance of the collector plate. Ordinarily, collectors are tested at clear sky conditions in which the solar radiation is nearly normal to the collector surface. For collectors facing south, tilted within 15° of the latitude, the ratio of the monthly average to normal incidence transmittance-absorptance product, $(\overline{\tau\alpha})/(\tau\alpha)_n$, during most of the heating season is 0.96 for a one cover and 0.94 for a two-cover collector. For collectors tilted more than 15° from the latitude, $(\overline{\tau\alpha})/(\tau\alpha)_n$ can be evaluated using the equations given in Section 3.5. $(\overline{\tau\alpha})/(\tau\alpha)_n$ is needed in Chapter 5 to evaluate long-term solar heating system performance.

CHAPTER 4
HEATING LOADS

4.1 HEATING LOAD CALCULATIONS

Unlike conventional heating systems using gas, oil, or electricity, both the efficiency and the economics of solar heating depend greatly upon the size of the solar heating system in relation to the size of the space and/or water heating loads. For conventional systems, it is ordinarily sufficient to estimate the design heating load (i.e., the maximum probable heating load) in order to size the heating equipment. In contrast, estimates of the long-term average heating load for each month are required to design solar heating systems.

A variety of factors influence heating loads, such as the geographic location of the building, its architectural design, orientation, construction quality, and the particular lifestyle of the occupants. Many different methods of calculating space heating loads have been developed, ranging in complexity from the simple degree-day method to detailed computer simulations using hourly meteorological data. All of these methods involve some degree of uncertainty.

In northern climates, the degree-day method of calculating space heating loads should be adequate to estimate the monthly average space heating loads needed to design solar heating systems. A brief review of the degree-day method and an example calculation are presented in the following section for the convenience of readers who are not familiar with the degree-day method and who do not have other preferred methods. If the reader has another way he prefers to calculate the monthly heating load for a building, he should use it; the method of estimating long-term system performance presented in the next chapter does not depend on the use of the degree-day concept.

4.2 THE DEGREE-DAY METHOD

The degree-day method of estimating the space heating load of a building is based upon the fact that the amount of heat required to maintain a comfortable indoor temperature is primarily dependent upon the difference between the indoor and outdoor temperatures. The monthly space heating load, L_S, for a building maintained at 22 C (72 F) is assumed to be proportional to the number of degree-days during the month.

$$L_S = UA \times DD \tag{4.1}$$

where

DD is the number of degree-days in a month

UA is the building overall energy loss coefficient-area product

The number of degree-days in a single day is the difference between 18.3 C (65 F) and the mean daily temperature (calculated as the average of the maximum and minimum daily temperatures). If the mean daily temperature is above 18.3 C, the number of degree-days is taken to be zero. Degree-days are calculated using 18.3 C, rather than 22 C because energy generation within the building (from the stove, lights, electrical appliances, people, etc.) and the solar energy gains through windows are usually sufficient to raise the indoor temperature from 18.3 C to the comfort level. The number of degree-days for a month is the sum of the daily degree-days. Extensive space heating fuel consumption records have shown that the monthly space heating load of a building is nearly proportional to the monthly degree-days calculated in this manner (ASHRAE Systems Handbook, Chapter 43 (1973)). Long-term averages of the degree-days for each month for many North American locations are tabulated in Appendix 2.

The building overall energy loss coefficient-area product, UA, can be determined in several ways. For existing structures in which records of conventional fuel requirements have been kept, UA can be calculated as the amount of energy required to heat the building for a given period (considering both the heating value of the fuel and the furnace efficiency) divided by the total number of degree-days occurring during that period.

$$UA = (N_F \times H_F \times \eta_F)/DD \qquad 4.2$$

where

N_F is the units of fuel consumed

H_F is the heating value of the fuel (See following table)

η_F is the average furnace efficiency which is ordinarily between 0.5 and 0.6 for gas and oil furnaces (Hise and Holman (1975)) and 1.0 for electrical heating

HEATING VALUE OF FUELS

NATURAL GAS	41.0 MJ/m^2	1100 BTU/ft^2
FUEL OIL	39.0 MJ/liter	140000 BTU/gal
ELECTRICITY	3.6 MJ/kW-hr	3400 BTU/kW-hr

For new structures, UA must be calculated from the details of the building construction. This is done by estimating the design heating load in the manner described in Chapter 21 of the ASHRAE Handbook of Fundamentals (1972) and dividing it by the design temperature difference.

$$UA = \frac{\text{DESIGN HEATING LOAD}}{\text{DESIGN TEMPERATURE DIFFERENCE}} \qquad 4.3$$

The calculation of the design heating load is rather tedious. However, this calculation need only be made once for any one building. The method of calculating the design heating load given in ASHRAE is straight forward, and illustrated by several examples. Additional information on design heat load calculations can be found in the text by Jennings (1970), in the Carrier System Design Manual (1972), and in the Insulation Manual for Homes and Apartments prepared by NAHB (1971). The method will not be repeated here. The following example demonstrates how the design heating load information can be used to estimate the average monthly space heating load of a building.

EXAMPLE 4.1 Estimation of Monthly Space Heating Loads in Madison

In Example 7 of Chapter 21 of the ASHRAE Handbook of Fundamentals (1972), the design heating load of a building located in Syracuse, New York is calculated to be 22042 W (7530 BTU/hr). The design temperature difference is 47.2 C (85 F). Estimate the average monthly space heating load of this building if it were located in Madison, Wisconsin.

The building overall energy loss coefficient-area product, UA, is the design heating load divided by the design temperature difference.

$$UA = (22042\ W)/(47.2\ C) = 467\ W/C$$

The product of UA and the number of degree-days for each month results in the average monthly space heating load. The long-term average number of degree-days for each month in Madison can be found in Appendix 2 in C-day units. Thus, for January, the space heating load, L_s, is

$$\begin{aligned} L_s &= 467\ W/C \times 839\ C\text{-days} \times 24\ hr/day \\ &\quad \times 3.6\ kJ/W\text{-}hr \\ &= 33.5 \times 10^9\ J \quad (33.5\ GJ) \end{aligned}$$

f-Chart Worksheet 1 has been prepared to organize the calculation of both the space and water heating loads. The space heating load for each month of the year appears in column C3 of the worksheet (Table 4.1). A blank worksheet is provided in Appendix 5.

TABLE 4.1 WORKSHEET FOR EXAMPLE 4.1

F-CHART WORKSHEET 1

HEATING LOADS

A. $UA = \frac{\text{Design Space Heating Load [W]}}{\text{Design Temperature Difference [°C]}} \; \frac{22042}{47.2} = \underline{467}$ [W/°C] (See Section 4.2)

B. Water Usage = 400 [liters/day] x 4190 x $(T_w - T_m)$[J/liter] = 82.1×10^6 [J/Day]

Month	C1. Days Per Mo.	C2. Heating Degree Days [°C-day] (from Appendix 2)	C3. Space Heating Load [J/Month] (86400)(A.)(C2.)	C4. Domestic Water Load [J/Month] (See Section 4.3) (B.)(C1.)	C5. Total Load [J/Month] (C3.)+(C4.)
Jan	31	830	33.5×10^9	2.55×10^9	36.1×10^9
Feb	28	696	28.1	2.30	30.4
Mar	31	599	24.2	2.55	26.7
Apr	30	328	13.2	2.46	15.7
May	31	165	6.7	2.55	9.3
Jun	30	40	1.6	2.46	4.1
Jul	31	8	0.3	2.55	2.8
Aug	31	22	0.9	2.55	3.5
Sep	30	96	3.9	2.46	6.4
Oct	31	263	10.6	2.55	13.2
Nov	30	505	20.4	2.46	22.9
Dec	31	742	29.9	2.55	32.5
Totals	365	4294	173.7×10^9	30.00×10^9	203.3×10^9

4.3 DOMESTIC WATER HEATING

The actual water heating load (i.e., the amount of energy required to heat water for domestic purposes) is highly dependent upon the lifestyle of the building occupants. On the average, however, each family member requires about 100 liters/day (25 gal/day) of domestic hot water. An average time distribution of water usage is shown in Figure 4.1.

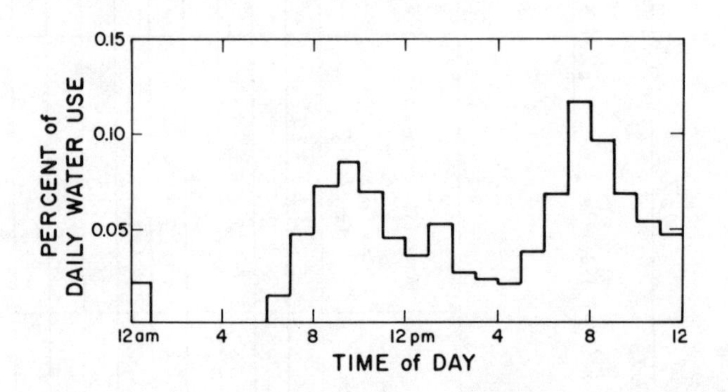

FIGURE 4.1
AVERAGE TIME DISTRIBUTION OF WATER USAGE

The monthly water heating load, L_w, can thus be estimated as

$$L_w = N \times (\# \text{ of persons}) \times 100 \times (T_w - T_m) \times \rho \times C_p \qquad 4.4$$

where

N is the number of days in the month

T_w is the minimum acceptable temperature for hot water; approximately 60 C (140 F)

T_m is the mains supply water temperature

ρ is the density of water (1 kg/liter or 8.33 lbm/gal)

C_p is the specific heat of water (4190 J/kg-C

or 1 BTU/lbm-F)

Note that the energy needs given by Equation 4.4 can be adjusted upward or downward if higher or lower hot water needs per person are anticipated.

EXAMPLE 4.2 Estimation of Monthly Domestic Water Heating Loads

Estimate the average water heating load each month for a family of four in Madison, Wisconsin. The mains water temperature is 11 C (51 F).
From Equation 4.4,

$$L_w = 30 \text{ days} \times 4 \times 100 \text{ l/day} \times (60-11) \text{ C} \times 1 \text{ kg/l} \times 4190 \text{ J/kg-C}$$
$$= 2.46 \times 10^9 \text{ J} \quad (2.46 \text{ GJ})$$

The water heating load for each month appears in column C4 of worksheet 1 (Table 4.1).

The monthly total load, L, is the sum of the space and domestic water heating loads.

$$L = L_s + L_w \qquad 4.5$$

The total space and water heating loads for each month in Madison are tabulated in column C5 of Table 4.1.

4.4 INSTITUTIONAL WATER HEATING

For buildings other than residences, the same basic methods of estimating water heating loads can be used, i.e., the amount of energy needed is the product of the volume of hot water required, the density of water, the specific heat of water, and the difference between the desired hot water temperature and the mains supply water temperature. On a monthly basis,

$$L_w = V \times \rho \times C_p \times (T_w - T_m) \qquad 4.6$$

where V is the volume of hot water required for a month and other terms are as defined under Equation 4.4.

Hot water supplies for commercial buildings may be needed on five or six days of the week, and not on weekends. If water is not used for one or two days of the week, the temperature of the water in the storage tank will rise, and energy will then be collected less efficiently. In this case, the fraction of the heating load supplied by solar energy estimated by the methods presented in Chapter 5 will be too high.

4.5 SUMMARY

The amount of collector area required for a particular application depends strongly on monthly space and water heating loads. The calculation of the heating loads is often the most difficult of the calculations needed to design solar heating systems. Estimating the space heating load using the degree-day method described in Section 4.2 is straight forward once UA, the building overall loss coefficient-area product, is known. Unless other information is available, UA must be estimated as the design heating load divided by the design temperature difference, and the design heating load requires tedious calculations. Methods of determining the design heating load are not presented here since these can be found in a number of other references.

Estimates of the long-term monthly average heating loads are needed for the design procedures in Chapter 5. These procedures do not depend on the degree-day calculation method. Estimates of the heating loads obtained in some other way, such as by computerized calculations, can be used.

CHAPTER 5
LONG-TERM PERFORMANCE OF SOLAR HEATING SYSTEMS

5.1 THE f-CHART METHOD

One approach to the problem of determining economic solar heating system designs is to use computer simulations directly as a design tool. This application was one incentive for the development of the general simulation program, TRNSYS (Klein et al. 1973). However, the use of computer simulations to aid in the design of every solar heating application is not satisfactory for those architects, contractors, and heating engineers concerned with the design of small buildings who do not have access to computing facilities. Simulations will remain an important design tool for large and nonstandard systems, but the widespread utilization of solar heating will require a simplified design procedure for use by the heating industry, especially for standard types of systems where the cost of detailed simulations cannot be justified. For these reasons, the "f-chart" method has been developed and is described here.

Our approach is to identify the important dimensionless variables of solar heating systems and to use detailed computer simulations to develop correlations between these variables and the long-term performance of these systems. The correlations developed for the liquid and air heating systems and for domestic water heating systems are presented in graphical and equation form and referred to as the "f-charts". The "f-chart" method of estimating solar heating system performance has been compared with the few long-term system performance data available in 1977, and with detailed computer simulations as described in Klein (1976) and in Klein et al. (1976a,b).

The result is a simple method requiring only monthly average meteorological data which can be used to estimate the long-term thermal performance of solar heating systems as a function of the major system design parameters. Combined with costs and meteorological data for the location in question, the f-charts provide a method by which architects and heating engineers can easily determine the thermal performance of solar heating systems, and thus optimize solar heating system designs in the light of costs.

5.2 IDENTIFICATION OF DIMENSIONLESS SYSTEM VARIABLES

An overall energy balance on a solar heating system over a one month period can be written

$$Q_T - L + E = \Delta U \qquad 5.1$$

where

Q_T is the total useful solar energy delivered during the month

L is the sum of the space and water heating loads for the month

E is the total auxiliary energy required during the month

ΔU is the energy change in the storage unit

For the storage sizes commonly used in solar heating systems, ΔU for a month is small with respect to Q_T, L, and E and it can be considered to be zero. Equation 5.1 can then be rearranged so that

$$f = (L-E)/L = Q_T/L \qquad 5.2$$

where f is the fraction of the monthly total heating load supplied by solar energy.

Equation 5.2 cannot be used to calculate f directly since Q_T is a complicated function of the incident radiation, the ambient temperature, and the heating loads. However, by considering the parameters on which Q_T depends, as discussed in Chapter 2, Equation 5.2 suggests that f may be empirically related to the two dimensionless groups.

$$X = A F_R' U_L (T_{ref} - \bar{T}_a) \Delta t / L \qquad 5.3$$

$$Y = A F_R' (\overline{\tau\alpha}) \bar{H}_T N / L \qquad 5.4$$

where

- A is the area of the solar collector [m^2]
- F_R' is the collector-heat exchanger efficiency factor (Section 2.4)
- U_L is the collector overall energy loss coefficient [$W/C\text{-}m^2$]
- Δt is the total number of seconds in the month
- T_{ref} is a reference temperature determined to be 100 C
- $\bar{T}_a$ is the monthly average ambient temperature which is given for many locations in Appendix 2 [C]
- L is the monthly total heating load [J]
- $\bar{H}_T$ is the monthly average daily radiation incident on the collector surface per unit area (Sections 3.2, 3.3) [J/m^2]
- N is the number of days in a month
- $(\overline{\tau\alpha})$ is the monthly average transmittance-absorptance product (Sections 3.4, 3.5)

These dimensionless groups have some physical significance. Y is related to the ratio of the total energy absorbed on the collector plate surface to the total heating load during the month. X is related to the ratio of a reference collector energy loss to the the total heating load during the month.

The equations for X and Y can be rewritten in slightly modified form for convenience in calculations:

$$X = F_R U_L \times (F_R'/F_R) \times (T_{ref} - \bar{T}_a) \times \Delta t \times A/L \qquad 5.3A$$

$$Y = F_R(\tau\alpha)_n \times (F_R'/F_R) \times (\overline{\tau\alpha})/(\tau\alpha)_n \times \bar{H}_T \times A/L \qquad 5.4A$$

Note that F_RU_L and $F_R(\tau\alpha)_n$ are obtained from collector test results by the methods noted in Section 2.3, F_R'/F_R by the method of Section 2.4, and $(\overline{\tau\alpha})/(\tau\alpha)_n$ by the method of Section 3.4 or Section 3.5. $\overline{T}_a$ is obtained from Appendix 2 for the month and location desired. $\overline{H}_T$ is found from $\overline{H}$ and $\overline{R}$ by the methods of Sections 3.2 and 3.3. The monthly loads, L, are determined as noted in Chapter 4. Values of A, the collector area, are selected for the calculation. Thus, all of the terms in these two equations are readily determined from available information.

5.3 LIQUID-BASED SOLAR SPACE HEATING SYSTEMS

f, the fraction of the monthly total load supplied by the solar space and water heating system shown in Figure 1.1, is given as a function of X and Y, the dimensionless variables defined in Equations 5.3 and 5.4, in Figure 5.1. This correlation has been developed from the results of hundreds of detailed computer simulations for a number of locations for a large range of practical system design variables. The result, Figure 5.1, is referred to as the f-chart for liquid-based solar space heating systems.

To determine f, the fraction of the heating load supplied by solar energy for a month, values of X and Y are calculated for the collector and heating load in question. (As will be seen in the following examples, all of the numbers needed for the calculations are readily available.) The value of f is determined at the intersection of X and Y on the chart. For example, values of X and Y of 4.0 and 1.0 respectively indicate that f is 0.57. This is done for each month of the year. The solar energy contribution for the month is the product of f and the total heating load, L, for the month. The fraction of the annual heating load supplied by solar energy , $\mathcal{F}$, is then the sum of the monthly solar energy contributions divided by the annual load.

The relationship between X, Y, and f in Figure 5.1 can also be expressed in equation form:

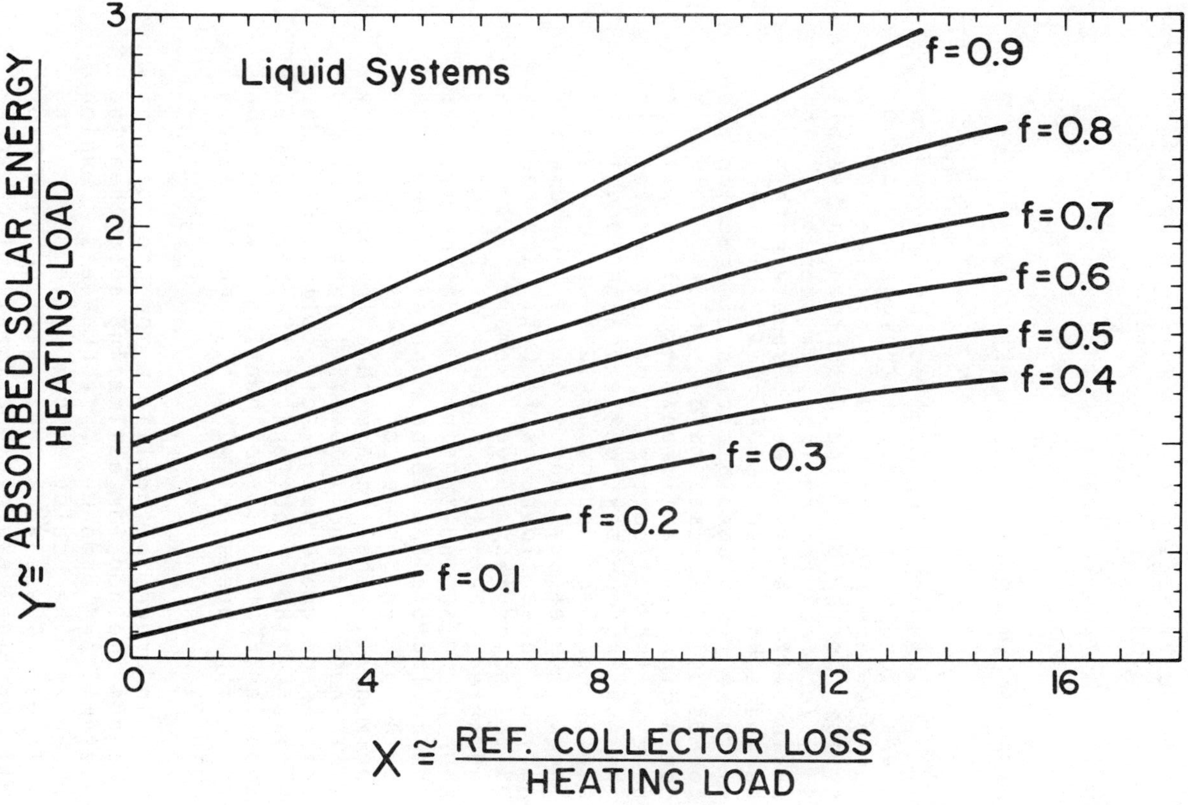

FIGURE 5.1
F-CHART FOR LIQUID SYSTEMS

$$f = 1.029Y - 0.065X - 0.245Y^2 + 0.0018X^2 + 0.0215Y^3 \quad 5.5$$

$$\text{for} \quad 0<Y<3 \quad \text{and} \quad 0<X<18$$

EXAMPLE 5.1 Performance of a Liquid-Based System

Liquid heating collectors having two glass covers and a flat-black absorber surface are to be used in a solar heating system for the building described in Example 4.1, located in Madison, Wisconsin. The system will be designed to supply both space heat and hot water for a family of four. To minimize the possibilities of freezing and corrosion, a separate flow circuit containing an antifreeze solution will be used in the collectors. Both flowrates in the collector-tank heat exchanger are to be 0.0139 kg/s per square meter of collector area. The effectiveness of the heat exchanger at these flowrates is 0.7. The results of collector tests (see Example 2.1) have indicated that the values of $F_R(\tau\alpha)_n$ and F_RU_L are 0.68 and 3.75 W/C-m^2, respectively. The collectors are to be mounted facing due south at an angle of 58° with respect to horizontal. The storage tank in the system will have a capacity of 75 liters of stored water per square meter of collector area. Estimate the fraction of the heating load supplied by the solar heating system for collector areas of 25, 50, and 100 square meters.

In order to facilitate the use of the f-chart method for estimating solar heating system performance, worksheets which organize the calculations have been developed. These worksheets will be used for all of the example problems. Blank worksheets are included in Appendix 5.

The first step is to determine the monthly total (space and domestic water) heating loads. Estimates of the monthly space heating load of the building were calculated in Example 4.1. The monthly domestic water heating load was calculated in Example 4.2. The total heating load for each month is the sum of the space and domestic water heating loads which is tabulated in column C5 of f-Chart Worksheet 1 (Table 4.1).

The fraction of the monthly total heating load supplied by solar energy, f, is a function of the dimensionless groups X and Y in Equations 5.3 and 5.4. X and Y must be calculated for each month and collector size in the following manner. Rearranging Equation 5.3,

$$X/A = F_R U_L \times (F_R'/F_R) \times (100-\bar{T}_a) \times \Delta t/L$$

For the collectors considered, $F_R U_L$ has been found to be 3.75 W/C-m^2 from the collector test results in Example 2.1. The collector-heat exchanger correction factor, (F_R'/F_R), has been calculated in Example 2.3 to be 0.97 for the conditions in this system. The product of $F_R U_L$ and (F_R'/F_R) appears as item C of f-Chart Worksheet 2 (Table 5.1). Monthly average temperatures for Madison can be found in Appendix 2. The difference between 100 C, a reference temperature, and the monthly average ambient temperature is tabulated in column C7 of worksheet 2. The number of seconds in a month, Δt, is given in column C6 . Monthly values of X/A are found as the product of $F_R' U_L$, $(100-\bar{T}_a)$, and Δt divided by the monthly total load. These results appear in column C8 (Table 5.1).

Rearranging Equation 5.4,

$$Y/A = F_R(\tau\alpha)_n \times (F_R'/F_R) \times (\overline{\tau\alpha})/(\tau\alpha)_n \times N \times \bar{H}_T/L$$

$F_R(\tau\alpha)_n$ has been determined to be 0.68 in Example 2.1 for the collectors considered in this problem. The collector-heat exchanger

TABLE 5.1 WORKSHEET FOR EXAMPLE 5.1

F-CHART WORKSHEET 2

ITEMS MAKING UP X AND Y

C. $F_R U_L (F'_R/F_R)$ = 3.64 [$W/m^2 {}^\circ C$] (See Sections 2.3 and 2.4)

D. $F_R(\tau\alpha)_n(F'_R/F_R)$ = 0.66 (See Sections 2.3 and 2.4)

	C6. Seconds Per Month	C7. $(100-\bar{T}_a)$[°C] ($\bar{T}_a$ can be found in Appendix 2	C8. X/A[$1/m^2$] $\frac{(C.)(C7.)(C6.)}{(C5.)}$	C9. $(\overline{\tau\alpha})/(\tau\alpha)_n$ (See Sections 3.4 and 3.5)	C10. Daily Radiation on Collector [J/m^2-Day] (See Section 3.2)	C11. Y/A[$1/m^2$] $\frac{(D.)(C9.)(C10.)(C1.)}{(C5.)}$
Jan	2.68×10^6	107	0.029	0.94	13.2×10^6	0.0070
Feb	2.42×10^6	106	0.031	0.94	14.8×10^6	0.0085
Mar	2.68×10^6	100	0.037	0.93	17.4×10^6	0.0124
Apr	2.59×10^6	93	0.055	0.91	15.4×10^6	0.0177
May	2.68×10^6	87	0.091	0.90	15.3×10^6	0.0303
Jun	2.59×10^6	81	0.186	0.89	16.4×10^6	0.0705
Jul	2.68×10^6	79	0.275	0.90	16.7×10^6	0.1098
Aug	2.68×10^6	80	0.223	0.90	17.0×10^6	0.0914
Sep	2.59×10^6	85	0.125	0.92	18.4×10^6	0.0524
Oct	2.68×10^6	90	0.067	0.93	17.0×10^6	0.0245
Nov	2.59×10^6	99	0.041	0.94	11.6×10^6	0.0094
Dec	2.68×10^6	105	0.032	0.94	12.4×10^6	0.0073

correction factor, (F_R'/F_R), is 0.97. The product of $F_R(\tau\alpha)_n$ and (F_R'/F_R) is 0.66; this value is entered as item D of worksheet 2 (Table 5.1).

The ratios of the monthly average transmittance-absorptance product to that at normal incidence, $(\overline{\tau\alpha})/(\tau\alpha)_n$, for the collectors considered here tilted at an angle of 58° have been calculated in Example 3.3; these values appear in column C9.

The monthly average daily radiation on tilted surfaces can be estimated in the manner discussed in Sections 3.2 and 3.3. The monthly average daily radiation on a 58° surface in Madison, $\bar{H}_T$, has been calculated in Example 3.1, and the results appear in column C10 of worksheet 2. Monthly values of Y/A are calculated as the product of $F_R(\tau\alpha)_n$, $(\overline{\tau\alpha})/(\tau\alpha)_n$, N, and $\bar{H}_T$ divided by the monthly total load. The values of Y/A are tabulated in column C11.

There are three correction factors on f-Chart Worksheet 3 (Table 5.2), items E, F, and G. In this example, the values of the storage and load heat exchanger size have been chosen such that items E and F are one. The air flowrate correction factor does not apply to liquid systems. Each of these factors will be dealt with in subsequent sections, and examples of their use will be shown.

The values of X/A and Y/A in columns C12 and C13 are multiplied by the collector area to yield values of X and Y for each month and collector size considered. The values of X and Y are tabulated in columns C14 and C15 for each collector size. Monthly average values of f, the load fraction supplied by solar energy, are found as a function of X and Y from Figure 5.1 or Equation 5.5. The monthly load fractions are tabulated in columns C16. The solar energy delivered each month is the product of the load fraction and the monthly total load. These values are tabulated in column C17 for each collector size.

TABLE 5.2 WORKSHEET FOR EXAMPLE 5.1

F-CHART WORKSHEET 3

SOLAR HEATING LOAD FRACTION

E. Storage size correction factor (X/X_0) = 1.0

F. Load heat exchanger correction factor (Y/Y_0) = 1.0 (=1.0 for air systems)

G. Collector air flow rate correction factor (X/X_0) = 1.0 (=1.0 for liquid systems)

	C12. Corrected X/A (C8.)(E.)(G.)	C13. Corrected Y/A (C11.)(F.)	Area = 25 m² C14. X	C15. Y	C16. f	C17. (C16.)(C5.)	Area = 50 m² C14. X	C15. Y	C16. f	C17. (C16.)(C5.)	Area = 100 m² C14. X	C15. Y	C16. f	C17. (C16.)C5.)
Jan	0.029	0.0070	0.72	0.18	0.13	4.69×10^9	1.45	0.35	0.24	8.66×10^9	2.9	0.70	0.44	15.88×10^9
Feb	0.031	0.0085	0.78	0.21	0.16	4.86	1.55	0.43	0.30	9.12	3.1	0.85	0.53	16.10
Mar	0.037	0.0124	0.93	0.31	0.24	6.41	1.85	0.62	0.44	11.75	3.7	1.24	0.73	19.49
Apr	0.055	0.0177	1.38	0.44	0.32	6.02	2.75	0.89	0.57	8.95	5.5	1.77	0.87	13.70
May	0.091	0.0303	2.28	0.76	0.52	4.85	4.55	1.52	0.82	7.63	9.1	3.03	1.00	9.30
Jun	0.186	0.0705	4.65	1.78	0.92	3.77	9.30	3.57	1.00	4.10	18.6	7.05	1.00	4.10
Jul	0.275	0.1098	6.88	2.75	1.00	2.80	13.8	6.50	1.00	2.80	27.5	11.0	1.00	2.80
Aug	0.223	0.0914	5.58	2.29	1.00	3.50	11.2	4.57	1.00	3.50	22.3	9.14	1.00	3.50
Sep	0.125	0.0524	3.13	1.32	0.80	5.12	6.25	2.65	1.00	6.40	12.5	5.24	1.00	6.40
Oct	0.067	0.0245	1.68	0.61	0.44	5.80	3.35	1.23	0.74	9.77	6.7	2.45	1.00	13.20
Nov	0.041	0.0094	1.03	0.24	0.16	3.66	2.05	0.47	0.31	7.10	4.1	0.94	0.53	12.10
Dec	0.032	0.0073	0.80	0.18	0.13	4.23	1.60	0.37	0.25	7.80	3.2	0.73	0.45	14.63
					Totals	54.7×10^9				87.6×10^9				131.2×10^9

Annual Fractions by Solar (Total, C17.)/(Total, C5.) = 0.27 = 0.43 = 0.64

The sum of the solar energy delivered for the year divided by the annual heating load results in the annual fraction of the heating load supplied by solar energy. The annual load fraction is plotted as a function of collector size for this example in Figure 5.2. At zero collector area, of course, the fraction supplied by solar energy is zero. Information of this type will be used in a study of the economics of solar heating in the next chapter.

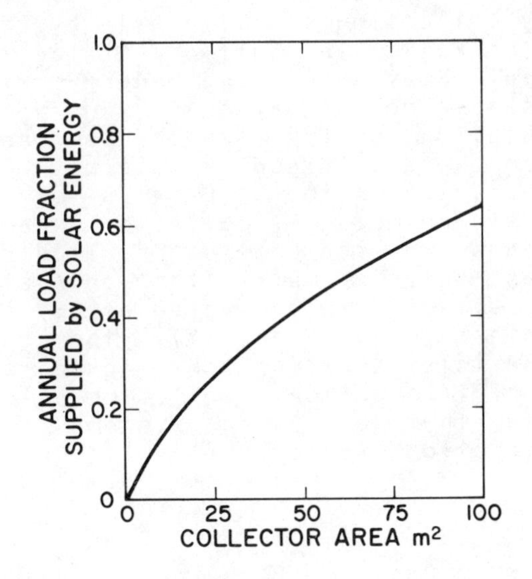

FIGURE 5.2
ANNUAL LOAD FRACTION SUPPLIED BY SOLAR ENERGY

5.3-1 COLLECTOR LIQUID FLUID FLOWRATE

Three system design parameters were held at fixed values to generate the f-chart. These are the collector fluid flowrate per unit collector area, the storage capacity per unit collector area, and the size of the load heat exchanger relative to the size of the space heating load. The effects of changes in these design parameters are considered here.

The optimum collector liquid fluid flowrate (i.e., the flowrate at which energy collection is maximized) is infinitely large. However, the dependence of sytem performance on the collector flowrate is asymptotic; only a small increase in the collector heat removal efficiency factor, F_R (and thus only a small gain in energy collection) is possible if the collector fluid capacitance rate (flowrate x specific heat) is increased beyond about 50 W/C per square meter of collector area which corresponds to an antifreeze solution flowrate of about 0.015 l/s-m^2 (0.022 gpm/ft^2). Low collector fluid flowrates can reduce energy collection significantly by reducing the value of F_R (or F_R'). In addition, if the flowrate is low, the fluid may boil and energy will be lost through a pressure relief valve.

The simulation results and correlations presented in this section were obtained using a collector fluid flowrate equivalent to 0.015 l/s of antifreeze solution per square meter of collector area. However, since a change in the collector liquid flowrate generally has only a small effect on system performance (being reflected in the value of F_R, and thus in the dimensionless groups X and Y), the correlations presented in this section for liquid systems are applicable for all practical collector liquid flowrates. (Note that air-based systems do require a flowrate correction factor, as will be shown in Section 5.4-1.)

5.3-2 STORAGE CAPACITY

Many simulation studies have been done to assess the effect of storage capacity on long-term system performance. It is found, all else being the same, that if storage capacity is greater than about 50 liters of water per square meter of collector, only small improvements in the yearly performance result from added storage capacity. When the costs of storage are considered, it appears that there are broad optima in the range of 50 to 100 liters of water per square meter of collector.

The f-chart (Equation 5.5 or Figure 5.1) has been generated for a storage capacity of 75 liters of stored water per square meter of collector area. The f-chart can be used to estimate the performance of systems with other storage capacities by modifying the dimensionless group X by the storage size correction factor given in Figure 5.3 or Equation 5.6.

$$\text{STORAGE SIZE CORRECTION FACTOR} = (X_c/X) = (M/75)^{-0.25} \qquad 5.6$$

$$\text{for} \quad 37.5 < M < 300$$

where M is the storage capacity in liters of water per square meter of collector area.

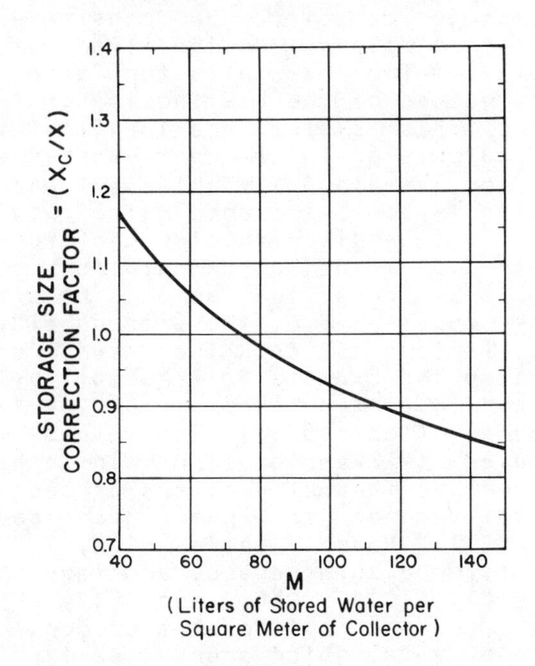

FIGURE 5.3
STORAGE SIZE CORRECTION FACTOR

EXAMPLE 5.2 Effect of Storage Capacity

The engineer responsible for the design of the solar heating system in Example 5.1 speculates that the system performance could be significantly improved if the storage capacity were larger. Estimate the fraction of the heating load supplied by the solar heating system in Example 5.1 if the storage capacity were doubled.

The storage capacity for the solar heating system in Example 5.1 was 75 liters of water per square meter of collector area. The storage capacity now considered is double this figure or 150 liters of water per square meter of collector area. The monthly values of the heating load, X/A, and Y/A for this system are identical to the values calculated in f-Chart Worksheets 1 and 2 for Example 5.1 (Tables 4.1 and 5.1). The effects of storage size on the calculations begin with item E of worksheet 3 (Table 5.3) with the calculation of the storage size correction factor. From Figure 5.3 or Equation 5.6, the storage size correction factor is 0.84. The values of X/A calculated for Example 5.1 are multiplied by 0.84 and tabulated in column C12 of f-Chart Worksheet 3 (Table 5.3). The values of Y/A calculated in Example 5.1 remain unchanged. From here on, the calculations proceed in an identical manner to that described in Example 5.1. Monthly values of X, Y, f, and the delivered solar energy are tabulated in columns C14, C15, C16, and C17, respectively, for each collector area considered. The annual total solar energy delivery and the fraction of the annual total load supplied by solar energy appear below column 17 for each collector area.

By comparing the annual load fractions calculated in this example with those calculated in Example 5.1, it can be seen that the increase in annual energy delivery resulting from doubling the storage capacity

TABLE 5.3 WORKSHEET FOR EXAMPLE 5.2

F-CHART WORKSHEET 3

SOLAR HEATING LOAD FRACTION

E. Storage size correction factor (X/X_0) = 0.84

F. Load heat exchanger correction factor (Y/Y_0) = 1.0 (=1.0 for air systems)

G. Collector air flow rate correction factor (X/X_0) = – (=1.0 for liquid systems)

	C12. Corrected X/A (C8.)(E.)(G.)	C13. Corrected Y/A (C11.)(F.)	Area = 25 m² C14. X	C15. Y	C16. f	C17. (C16.)(C5.)	Area = 50 m² C14. X	C15. Y	C16. f	C17. (C16.)(C5.)	Area = 100 m² C14. X	C15. Y	C16. f	C17. (C16.)C5.)
Jan	0.024		0.61	0.18	0.13	4.69×10⁹	1.21	0.35	0.26	9.39×10⁹	2.42	0.70	0.46	16.68×10⁹
Feb	0.026		0.66	0.21	0.17	5.17	1.31	0.43	0.31	9.42	2.62	0.85	0.55	16.72
Mar	0.031	Same As	0.78	0.31	0.25	6.68	1.56	0.62	0.45	12.02	3.13	1.24	0.76	20.29
Apr	0.046		1.16	0.44	0.34	6.34	2.32	0.89	0.59	9.26	4.64	1.77	0.91	14.29
May	0.077	Column C13	1.92	0.76	0.53	4.93	3.83	1.52	0.85	7.91	7.67	3.03	1.00	9.30
Jun	0.156		3.91	1.78	0.95	3.90	7.82	3.57	1.00	4.10	15.6	7.13	1.00	4.10
Jul	0.231	of	5.79	2.75	1.00	2.80	11.6	5.50	1.00	2.80	23.1	11.0	1.00	2.80
Aug	0.188		4.69	2.29	1.00	3.50	9.38	4.57	1.00	3.50	18.8	9.14	1.00	3.50
Sep	0.105	Table 5.2	2.63	1.32	0.82	5.25	6.26	2.65	1.00	6.40	10.5	5.29	1.00	6.40
Oct	0.056		1.41	0.61	0.46	6.07	2.83	1.23	0.76	10.03	6.65	2.45	1.00	13.20
Nov	0.034		0.87	0.24	0.17	3.89	1.73	0.47	0.33	7.56	3.46	0.94	0.57	13.05
Dec	0.027		0.67	0.18	0.14	4.55	1.35	0.37	0.26	8.45	2.69	0.73	0.47	15.28
					Totals	56.8×10⁹				90.8×10⁹				135.6×10⁹

Annual Fractions by Solar
(Total, C17.)/(Total, C5.) = 0.28 = 0.45 = 0.67

is small, particularily for the smaller collector areas.

5.3-3 LOAD HEAT EXCHANGER SIZE

The size of the load heat exchanger can significantly affect the performance of the solar heating system. When the heat exchanger used to heat the building air is reduced in size, the storage tank temperature must be increased to supply the same amount of heat. This results in higher collector fluid inlet temperatures which reduces the collector efficiency. A measure of the size heat exchanger needed for a specific building is provided by the dimensionless parameter, $\varepsilon_L C_{min}/UA$. Here, ε_L is the effectiveness of the water-air load heat exchanger. (See Appendix 1 for an explanation and example calculation of heat exchanger effectiveness.) C_{min} is the minimum fluid capacitance rate (mass flowrate times the specific heat of the fluid) in the heat exchanger, and is generally that of the air for this heat exchanger. UA is the building overall energy loss coefficient-area product discussed in Chapter 4.

The optimum value of $\varepsilon_L C_{min}/UA$ from a thermal standpoint is infinitely large. However, system performance is asymptotically dependent upon the value of this parameter, and for values of $\varepsilon_L C_{min}/UA$ greater than 10, system performance will be nearly the same as that for the infinitely large value. The reduction in system performance due to a too small load heat exchanger will be appreciable for values of $\varepsilon_L C_{min}/UA$ less than about 1. Practical values of $\varepsilon_L C_{min}/UA$ are generally between 1 and 3 when the cost of the heat exchanger is considered.

The f-chart (Equation 5.5 or Figure 5.1) has been developed using a value of $\varepsilon_L C_{min}/UA$ equal to 2. The performance of systems having other values of $\varepsilon_L C_{min}/UA$ can be calculated from the f-chart by modifying the dimensionless group Y as indicated in Figure 5.4 or Equation 5.7.

LOAD HEAT EXCHANGER CORRECTION FACTOR $= (Y_c/Y)$

$$= (0.39+0.65\ \text{EXP}(-0.139/(\varepsilon_L C_{min}/UA)) \qquad 5.7$$

for $\quad 0.5 < \varepsilon_L C_{min}/UA < 50$

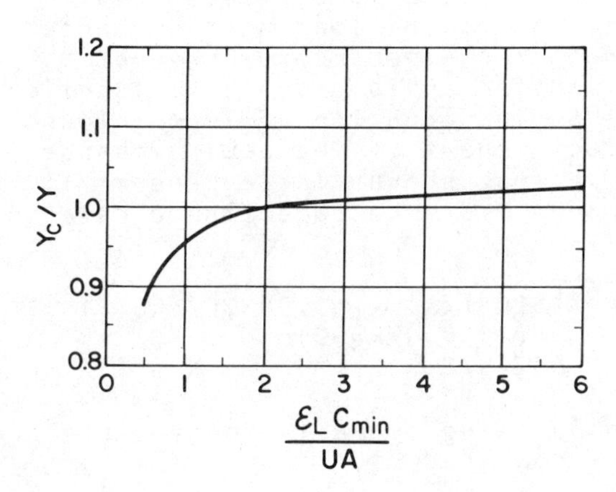

FIGURE 5.4
LOAD HEAT EXCHANGER SIZE CORRECTION FACTOR

EXAMPLE 5.3 Effect of Load Heat Exchanger

The effectiveness, ε_L, and minimum capacitance rate, C_{min}, of the crossflow heat exchanger which transfers heat from the storage tank to the building air were not specified in Example 5.1. As a result, it was assumed that the heat exchanger would be sized such that $\varepsilon_L C_{min}/UA=2$, the value used in the development of the f-chart. It is now known that the air flowrate in the heat exchanger will be 520 liters of standard air per second (1100 cfm), the water flowrate will be 0.694 l/s (11 gpm), and the effectiveness of the heat exchanger at these flowrates is 0.69, as determined in Example A1.1 in Appendix 1. Revise the calculations in Example 5.1, using this additional information.

The monthly values of the heating load, X/A, and Y/A in worksheets 1 and 2 for Example 5.1 (Tables 4.1 and 5.1) are not affected by the additional information concerning the load heat exchanger. Changes in the calculations begin with the calculation of the load heat exchanger correction factor, item F of worksheet 3. This correction factor is given as a function of $\varepsilon_L C_{min}/UA$ in Equation 5.7 or Figure 5.4. The effectiveness of the heat exchanger, ε_L, is 0.69. As shown in Appendix 1, C_{min} corresponds to the capacitance rate of the air.

$$
\begin{aligned}
C_{min} &= (520\ \text{l/s}) \times (0.001204\ \text{kg/l}) \\
&\quad \times (1010\ \text{J/kg-C}) \\
&= 632\ \text{W/C}
\end{aligned}
$$

UA, the building overall energy loss coefficient-area product, is 467 W/C.

$$\varepsilon_L C_{min}/UA = 0.69 \times (632\ \text{W/C})/(467\ \text{W/C}) = 0.93$$

Therefore, from Equation 5.7 or Figure 5.4, the load heat exchanger correction factor is 0.95. The values of Y/A in column C11 of worksheet 2 (Table 5.1) must be multiplied by 0.95. The corrected values of Y/A appear in column C13 of worksheet 3 (Table 5.4). The values of X/A in column C12 are the same as those in column C8 of worksheet 2 (Table 5.2).

From here on, the calculations proceed as described in Example 5.1. Monthly values of X and Y are calculated for each collector area. The fractions of the monthly heating load supplied by solar energy are determined from Equation 5.5 or Figure 5.1 and tabulated in column C17 of Table 5.4. The monthly solar energy delivered in column C17 is the product of f and the monthly total heating load. The annual solar energy delivered and fraction of the annual load

TABLE 5.4 WORKSHEET FOR EXAMPLE 5.3

F-CHART WORKSHEET 3

SOLAR HEATING LOAD FRACTION

E. Storage size correction factor (X/X_o) = 1.0

F. Load heat exchanger correction factor (Y/Y_o) = 0.95 (=1.0 for air systems)

G. Collector air flow rate correction factor (X/X_o) = 1.0 (=1.0 for liquid systems)

	C12. Corrected X/A (C8.)(E.)(G.)	C13. Corrected Y/A (C11.)(F.)	Area = 25m² C14. X	C15. Y	C16. f	C17. (C16.)(C5.)	Area = 50m² C14. X	C15. Y	C16. f	C17. (C16.)(C5.)	Area = 100m² C14. X	C15. Y	C16. f	C17. (C16.)C5.)
Jan	0.029	0.0067	0.72	0.17	0.12	4.33x10⁹	1.45	0.34	0.23	8.30x10⁹	2.90	0.67	0.41	14.90x10⁹
Feb	0.031	0.0081	0.78	0.20	0.15	4.56	1.55	0.41	0.28	8.51	3.10	0.81	0.50	15.20
Mar	0.037	0.0118	0.93	0.30	0.22	5.87	1.85	0.59	0.41	10.95	3.70	1.18	0.69	18.42
Apr	0.055	0.0169	1.38	0.42	0.31	4.87	2.75	0.85	0.54	8.48	5.50	1.69	0.84	13.19
May	0.091	0.0289	2.28	0.72	0.49	4.56	4.55	1.45	0.78	7.25	9.10	2.89	1.00	9.30
Jun	0.186	0.0680	4.65	1.70	0.88	3.61	9.30	3.40	1.00	4.10	18.6	6.80	1.00	4.10
Jul	0.275	0.1047	6.88	2.62	1.00	2.80	13.8	5.24	1.00	2.80	27.5	10.5	1.00	2.80
Aug	0.223	0.0872	5.58	2.18	0.99	3.47	11.15	4.36	1.00	3.50	22.3	8.72	1.00	3.50
Sep	0.125	0.0504	3.13	1.26	0.77	4.93	6.25	2.52	1.00	6.40	12.5	5.04	1.00	6.40
Oct	0.067	0.0234	1.68	0.59	0.42	5.52	3.35	1.17	0.70	9.24	6.70	2.34	0.99	13.07
Nov	0.041	0.0090	1.03	0.23	0.15	3.44	2.05	0.45	0.29	6.64	4.10	0.90	0.50	11.45
Dec	0.032	0.0070	0.80	0.17	0.12	3.90	1.60	0.35	0.23	7.48	3.20	0.70	0.42	13.42
					Totals	51.9x10⁹				83.7x10⁹				125.8x10⁹

Annual Fractions by Solar (Total, C17.)/(Total, C5.) = 0.26 = 0.41 = 0.62

supplied by solar energy appear below column C17 for each collector area. As expected, the performance of the solar heating system estimated in this example is somewhat lower than that estimated in Example 5.1 because of the smaller load heat exchanger.

5.4 SOLAR AIR HEATING SYSTEMS

In a manner identical to that for the liquid-based systems, the correlation of f, the monthly heating load fraction supplied by the solar air heating system shown in Figure 1.2, to the dimensionless groups, X and Y, has been determined using computer simulations. The correlation is given in Figure 5.5 and in Equation 5.8. It is used in the same manner as the f-chart for liquid-based systems. The definitions of X and Y given in Equations 5.3 and 5.4 (and 5.3A and 5.4A) apply to both air and liquid systems.

$$f = 1.040Y - 0.065X - 0.159Y^2 + 0.00187X^2 - 0.0095Y^3 \qquad 5.8$$

$$\text{for} \quad 0<Y<3 \quad \text{and} \quad 0<X<18$$

EXAMPLE 5.4 Performance of an Air System

A solar air heating system is being considered for the building described in Example 4.1. The system will be designed to supply both heat and hot water for a family of four in Madison, Wisconsin. Air heaters having the characteristics determined in Example 2.2 will be mounted on the roof of the building facing due south at an angle of 58° with respect to the horizontal. The collector air flowrate is anticipated to be 10.1 liter per second per square meter of collector area. The pebble bed will be sized such that there is 0.25 cubic meters of pebbles per square meter of collector area. Estimate the energy delivered by this

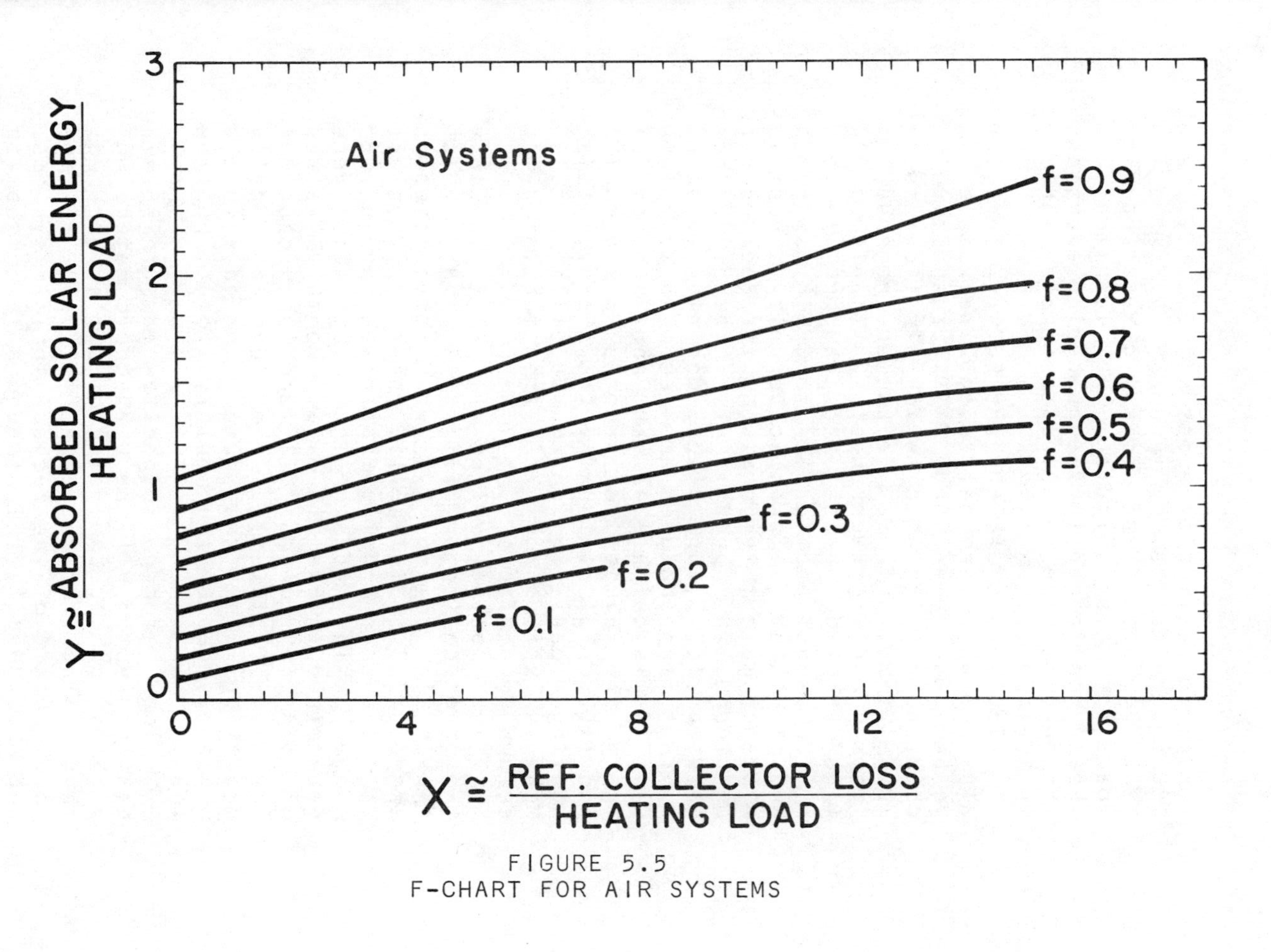

FIGURE 5.5
F-CHART FOR AIR SYSTEMS

system for collector areas of 25, 50, and 100 square meters.

The space and domestic water heating loads for this building are identical to those considered in the preceding examples. The heating load calculations are summarized in Table 4.1.

The values of $F_R U_L$ and $F_R(\tau\alpha)_n$ for the air heaters were determined in Example 2.2 to be 2.84 W/C-m^2 and 0.49, respectively, corresponding to a collector air flowrate of 10.1 l/s-m^2. These values are recorded as items C and D of worksheet 2 (Table 5.5). (The collector-heat exchanger factor, F_R'/F_R is not applicable for air systems, and its value is taken to be unity.) Monthly average ambient temperatures for Madison can be found in Appendix 2. The difference between 100 C and the monthly average ambient temperatures are recorded in column C7. X/A is calculated as the product of $F_R U_L$, the number of seconds in the month, and (100 C-$\bar{T}_a$) divided by the monthly total heating load (in column C5 of Table 4.1). Monthly values of X/A are tabulated in column C9 of Table 5.5.

The monthly values of $(\overline{\tau\alpha})/(\tau\alpha)_n$ for a two cover collector tilted at 58°, which appear in column C9, were calculated in Example 3.3. Similarily, the monthly values of the average daily radiation on a 58° surface in Madison in column C10 were determined in Example 3.1. Monthly values of Y/A in column C11 are the product of $F_R(\tau\alpha)_n$, $(\overline{\tau\alpha})/(\tau\alpha)_n$, the average daily radiation, and the number of days in the month divided by the monthly total heating load.

The storage capacity for this air heating system will be 0.25 cubic meters of pebbles per square meter of collector area. This is the storage capacity used in the development of the f-chart for air systems, and as a result, the storage size correction factor, item E of worksheet 3 (Table 5.6), is unity. (See Section 5.4-2.) Similarily,

TABLE 5.5 WORKSHEET FOR EXAMPLE 5.4

F-CHART WORKSHEET 2

ITEMS MAKING UP X AND Y

C. $F_R U_L (F'_R / F_R)$ = 2.84 $[W/m^2 °C]$ (See Sections 2.3 and 2.4)

D. $F_R (\tau\alpha)_n (F'_R / F_R)$ = 0.49 (See Sections 2.3 and 2.4)

	C6.	C7.	C8.	C9.	C10.	C11.
	Seconds Per Month	$(100-\bar{T}_a)$[°C] ($\bar{T}_a$ can be found in Appendix 2	X/A[$1/m^2$] $\frac{(C.)(C7.)(C6.)}{(C5.)}$	$(\overline{\tau\alpha})/(\tau\alpha)_n$ (See Sections 3.4 and 3.5)	Daily Radiation on Collector [J/m^2-Day] (See Section 3.2)	Y/A[$1/m^2$] $\frac{(D.)(C9.)(C10.)(C1.)}{(C5.)}$
Jan	2.68×10^6	107	0.023	0.94	13.2 $\times 10^6$	0.0052
Feb	2.42×10^6	106	0.024	0.94	14.8 $\times 10^6$	0.0063
Mar	2.68×10^6	100	0.029	0.93	17.4 $\times 10^6$	0.0092
Apr	2.59×10^6	93	0.044	0.91	15.4 $\times 10^6$	0.0131
May	2.68×10^6	87	0.071	0.90	15.3 $\times 10^6$	0.0226
Jun	2.59×10^6	81	0.145	0.89	16.4 $\times 10^6$	0.0523
Jul	2.68×10^6	79	0.215	0.90	16.7 $\times 10^6$	0.0815
Aug	2.68×10^6	80	0.174	0.90	17.0 $\times 10^6$	0.0664
Sep	2.59×10^6	85	0.098	0.92	18.4 $\times 10^6$	0.0389
Oct	2.68×10^6	90	0.052	0.93	17.0 $\times 10^6$	0.0182
Nov	2.59×10^6	99	0.032	0.94	11.6 $\times 10^6$	0.0070
Dec	2.68×10^6	105	0.025	0.94	12.4 $\times 10^6$	0.0054

TABLE 5.6 WORKSHEET FOR EXAMPLE 5.4

F-CHART WORKSHEET 3

SOLAR HEATING LOAD FRACTION

E. Storage size correction factor (X/X_o) = 1.0

F. Load heat exchanger correction factor (Y/Y_o) = 1.0 (=1.0 for air systems)

G. Collector air flow rate correction factor (X/X_o) = 1.0 (=1.0 for liquid systems)

	C12. Corrected X/A (C8.)(E.)(G.)	C13. Corrected Y/A (C11.)(F.)	Area = 25 m² C14. X	C15. Y	C16. f	C17. (C16.)(C5.)	Area = 50 m² C14. X	C15. Y	C16. f	C17. (C16.)(C5.)	Area = 100 m² C14. X	C15. Y	C16. f	C17. (C16.)C5.)
Jan			0.58	0.13	0.10	3.61 × 10⁹	1.15	0.26	0.19	6.86 × 10⁹	2.30	0.52	0.36	13.0 × 10⁹
Feb			0.60	0.16	0.12	3.65	1.20	0.32	0.24	7.30	2.40	0.63	0.44	13.4
Mar	Same as	Same as	0.73	0.23	0.18	4.81	1.45	0.46	0.35	9.35	2.90	0.92	0.64	17.1
Apr			1.10	0.33	0.25	3.93	2.20	0.66	0.48	7.54	4.40	1.31	0.82	12.9
May	Column C8	Column C11	1.78	0.57	0.43	4.00	3.55	1.13	0.75	6.98	7.10	2.26	1.00	9.30
Jun			3.63	1.31	0.86	3.53	7.25	2.62	1.00	4.10	14.5	5.23	1.00	4.10
Jul	of	of	5.38	2.04	1.00	2.80	10.7	4.08	1.00	2.80	21.5	8.15	1.00	2.80
Aug			4.35	1.66	1.00	3.50	8.70	3.32	1.00	3.50	17.4	6.64	1.00	3.50
Sep	Table 5.5	Table 5.5	2.45	0.97	0.71	4.54	4.90	1.95	1.00	6.40	9.80	3.89	1.00	6.40
Oct			1.30	0.46	0.36	4.75	2.60	0.91	0.65	8.58	5.20	1.82	1.00	13.20
Nov			0.80	0.18	0.13	2.98	1.60	0.35	0.24	5.50	3.20	0.70	0.46	10.50
Dec			0.63	0.14	0.10	3.25	1.25	0.27	0.19	6.18	2.50	0.54	0.36	11.70
					Totals	45.1 × 10⁹				75.1 × 10⁹				117.9 × 10⁹

Annual Fractions by Solar
(Total, C17.)/(Total, C5.) = 0.22 = 0.37 = 0.58

the collector air flowrate is to be 10.1 $l/s\text{-}m^2$ for which the collector air flowrate correction factor is also unity. (See Section 5.4-1.) Thus, the corrected values of X/A and Y/A in columns C12 and C13 are the same as the values in columns C8 and C11.

The values of X/A and Y/A are multiplied by the collector area to yield values of X and Y (in columns C14 and C15) for each month and collector area considered. Monthly values of f, the load fraction supplied by solar energy, are obtained from either Equation 5.8 or Figure 5.5 and recorded in column C16 of Table 5.6. The total solar energy delivered each month in column C17 is the product of f and the monthly total heating load. The annual total delivered energy divided by the annual total heating load results in the annual load fraction supplied by solar energy for each collector size.

5.4-1 COLLECTOR AIR FLOWRATE

The collector air flowrate and the storage capacity of the packed bed relative to the collector area were not varied in generating the f-chart. The effects of changes in these parameters are considered here.

The collector heat removal efficiency factor, F_R, which appears in the dimensionless variables X and Y, is a function of the collector air flowrate. Because of the cost of power for blowing air through the collectors, the capacitance rate in air heaters is ordinarily much lower than that in liquid heaters. As a result, air heaters generally have a lower value of F_R. Values of F_R (and thus $F_R(\tau\alpha)_n$ and F_RU_L) corresponding to the actual air flowrate in the collector must be used in calculating X and Y.

Aside from affecting the value of F_R, a change in collector air flowrate affects the thermal stratification in the pebble bed. An increase in air flowrate tends to improve system performance by increasing the value of F_R, but it also tends to decrease performance somewhat by reducing the degree of thermal stratification.

The f-chart for air heating systems (Equation 5.8 or Figure 5.5) was generated for a collector air flowrate of 10.1 l/s of air per square meter of collector area (2 cfm per square foot of collector area). The performance of systems having other collector air flowrates can be estimated by using the appropriate values of F_R, X, and Y, and then modifying the value of X as indicated in Figure 5.6 or Equation 5.9 to account for the degree of stratification in the pebble bed.

$$\text{COLLECTOR AIR FLOWRATE CORRECTION FACTOR} = (X_c/X) = (m/10.1)^{0.28} \qquad 5.9$$

$$\text{for} \quad 5 < m < 20$$

where m is the air flowrate in l/s per square meter of collector area.

EXAMPLE 5.5 Effect of Air Flowrate

A larger-capacity blower is to be used in the air heating system considered in Example 5.4. The collector air flowrate will be 15.1 l/s per square meter of collector area. Collector tests at this flowrate result in values of $F_R U_L$ and $F_R(\tau\alpha)_n$ equal to 3.01 $W/C\text{-}m^2$, and 0.52 respectively. Estimate the annual energy delivery for the system at this collector air flowrate.

The collector heat removal efficiency factor, F_R, is a function of the collector air flowrate. The collector tests have shown that increasing the air flowrate from 10.1 to 15.1 $l/s\text{-}m^2$ has increased F_R and thus $F_R U_L$ and $F_R(\tau\alpha)_n$ by 6%. As a result, the values of X/A and Y/A for this example (Table 5.7) are 6% larger than the values obtained on Table 5.5 in the preceding example.

The storage size correction factor, item E of worksheet 3 (Table 5.8), is unity since the storage capacity is to be 0.25

TABLE 5.7 WORKSHEET FOR EXAMPLE 5.5

F-CHART WORKSHEET 2

ITEMS MAKING UP X AND Y

C. $F_R U_L (F'_R/F_R)$ = 3.01 [$W/m^2 °C$] (See Sections 2.3 and 2.4)

D. $F_R (\tau\alpha)_n (F'_R/F_R)$ = 0.52 (See Sections 2.3 and 2.4)

	C6.	C7.	C8.	C9.	C10.	C11.
	Seconds Per Month	$(100-\bar{T}_a)$[°C] ($\bar{T}_a$ can be found in Appendix 2	X/A[$1/m^2$] $\frac{(C.)(C7.)(C6.)}{(C5.)}$	$(\overline{\tau\alpha})/(\tau\alpha)_n$ (See Sections 3.4 and 3.5)	Daily Radiation on Collector [J/m^2-Day] (See Section 3.2)	Y/A[$1/m^2$] $\frac{(D.)(C9.)(C10.)(C1.)}{(C5.)}$
Jan	2.68×10^6	107	0.024	0.94	13.2 $\times 10^6$	0.0055
Feb	2.42×10^6	106	0.025	0.94	14.8 $\times 10^6$	0.0067
Mar	2.68×10^6	100	0.031	0.93	17.4 $\times 10^6$	0.0098
Apr	2.59×10^6	93	0.044	0.91	15.4 $\times 10^6$	0.0139
May	2.68×10^6	87	0.075	0.90	15.3 $\times 10^6$	0.0240
Jun	2.59×10^6	81	0.154	0.89	16.4 $\times 10^6$	0.0554
Jul	2.68×10^6	79	0.228	0.90	16.7 $\times 10^6$	0.0864
Aug	2.68×10^6	80	0.184	0.90	17.0 $\times 10^6$	0.0704
Sep	2.59×10^6	85	0.104	0.92	18.4 $\times 10^6$	0.0412
Oct	2.68×10^6	90	0.055	0.93	17.0 $\times 10^6$	0.0193
Nov	2.59×10^6	99	0.034	0.94	11.6 $\times 10^6$	0.0074
Dec	2.68×10^6	105	0.027	0.94	12.4 $\times 10^6$	0.0057

TABLE 5.8 WORKSHEET FOR EXAMPLE 5.5

F-CHART WORKSHEET 3

SOLAR HEATING LOAD FRACTION

E. Storage size correction factor (X/X_o) = 1.0

F. Load heat exchanger correction factor (Y/Y_o) = 1.0 (=1.0 for air systems)

G. Collector air flow rate correction factor (X/X_o) = 1.12 (=1.0 for liquid systems)

	C12. Corrected X/A (C8.)(E.)(G.)	C13. Corrected Y/A (C11.)(F.)	Area = 25 m²				Area = 50 m²				Area = 100 m²			
			C14.	C15.	C16.	C17.	C14.	C15.	C16.	C17.	C14.	C15.	C16.	C17.
			X	Y	f	(C16.)(C5.)	X	Y	f	(C16.)(C5.)	X	Y	f	(C16.)C5.)
Jan	0.027		0.67	0.14	0.10	3.61×10^9	1.35	0.28	0.19	6.86×10^9	2.70	0.55	0.36	13.0×10^9
Feb	0.028	Same as	0.70	0.17	0.13	3.95	1.40	0.34	0.24	7.30	2.80	0.67	0.46	14.0
Mar	0.035		0.88	0.25	0.19	5.07	1.75	0.49	0.36	9.61	3.50	0.98	0.65	17.4
Apr	0.049	Column C11	1.23	0.35	0.26	4.08	2.45	0.70	0.49	7.69	4.90	1.39	0.84	13.2
May	0.084		2.10	0.60	0.44	4.09	4.20	1.20	0.76	7.07	8.40	2.40	1.00	9.3
Jun	0.173	of	4.33	1.38	0.86	3.54	8.65	2.77	1.00	4.1	17.3	5.54	1.00	4.1
Jul	0.255		6.38	2.16	1.00	2.80	12.75	4.32	1.00	2.8	25.5	8.64	1.00	2.8
Aug	0.205	Table 5.7	5.13	1.76	1.00	3.50	10.25	3.52	1.00	3.5	20.5	7.04	1.00	3.5
Sep	0.117		2.93	1.03	0.73	4.67	5.85	2.06	1.00	6.4	11.7	4.12	1.00	6.4
Oct	0.062		1.55	0.48	0.37	4.88	3.10	0.97	0.66	8.71	6.20	1.93	1.00	13.2
Nov	0.038		0.95	0.19	0.13	2.98	1.90	0.37	0.25	5.73	3.80	0.74	0.46	10.5
Dec	0.030		0.75	0.14	0.10	3.25	1.50	0.29	0.19	6.18	3.00	0.57	0.36	11.7
					Totals	46.4×10^9				75.9×10^9				119.1×10^9

Annual Fractions by Solar (Total, C17.)/(Total, C5.) = 0.23 = 0.37 = 0.59

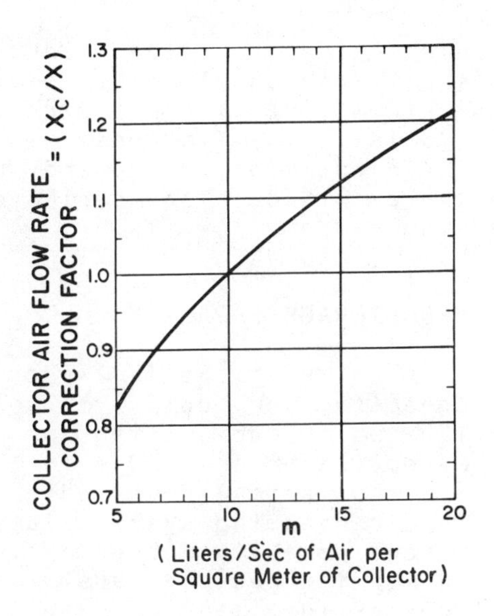

FIGURE 5.6
COLLECTOR AIR FLOWRATE CORRECTION FACTOR

cubic meters of pebbles per square meter of collector area, the storage capacity used to develop the f-chart correlation. (See Section 5.4-2.) The collector air flowrate correction factor with m equal to 15.1 $l/s\text{-}m^2$ is 1.12 from Figure 5.6 or Equation 5.9. The values of X/A in column C8 must be multiplied by 1.12. The corrected values of X/A appear in column C12 of Table 5.8. The values of Y/A are unaffected. From here on the calculations proceed as before. Monthly values of X, Y, f, and delivered solar energy appear in columns C14, C15, C16, and C17 respectively for each collector area.

It is interesting to note that the annual system performance was negligibly improved by increasing the collector air flowrate from 10.1 to 15.1 $l/s\text{-}m^2$ in this case. Although the collector heat removal efficiency factor was 6% higher at the larger air flowrate, the reduction of thermal stratification in the pebble bed resulting from the larger air flowrate tended to offset the increase in energy delivery. (In addition, the energy required to pump the air through the system would be higher at the higher flowrate.)

5.4-2 PEBBLE BED STORAGE CAPACITY

The results of many simulations in which the pebble bed storage capacity per unit collector area was varied from 0.125 to 1.0 cubic meter of pebbles per square meter of collector area (0.4 to 3.3 cubic feet per square foot of collector) have indicated that the performance of air heating systems is slightly less sensitive to storage capacity than are liquid systems. (One explanation for the reduced sensitivity is that the air heating system can operate in the collector-load mode, in which the storage component is not used. Another is that pebble beds are highly stratified; additional capacity is effectively added to the "cold" end of the bed, which is seldom heated and cooled to the same extent as the "hot" end.)

The f-chart for air heating systems (Equation 5.8 and Figure 5.5) was generated for a storage capacity of 0.25 cubic meters of pebbles per square meter of collector area. The performance of systems with other storage capacities can be determined by modifying the dimensionless group X as indicated in Figure 5.7 or Equation 5.10.

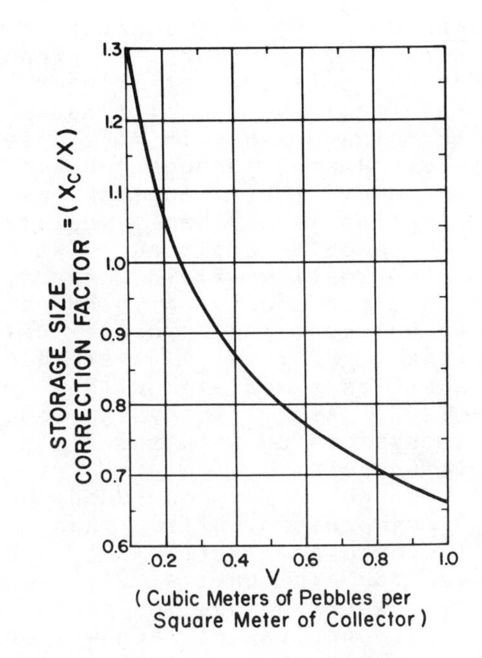

FIGURE 5.7
STORAGE SIZE CORRECTION FACTOR

$$\text{STORAGE SIZE CORRECTION FACTOR} = (X_c/X) = (V/0.25)^{-0.3} \qquad 5.10$$

$$\text{for} \quad 0.125 < V < 1.0$$

where V is the storage capacity in cubic meters of pebbles per square meter of collector area.

5.5 COMPARISON OF LIQUID AND AIR SYSTEMS

A comparison of the f-charts for the liquid and air systems, Figures 5.1 and 5.5, indicates that, for the same values of X and Y, the air system outperforms the liquid system, particularly for systems designed to supply a large fraction of the heating load. There are several reasons for this behavior. The average collector fluid inlet temperature is lower for the air system (and thus the collector efficiency is higher) than that for the liquid system at times when there is a space heating load, since in this case, room temperature air is circulated through the air heater and returned to the building. Another reason is that thermal stratification is ordinarily maintained at a higher level in pebble beds than in water tanks, in part because of the smaller fluid capacitance rates normally used in air heaters. A third reason is that a heat exchanger between the storage unit and the heating load is not required in an air heating system and thus the penalties associated with a load heat exchanger are avoided. Also, air system do not "dump" energy as liquid systems do when the fluid temperature reaches its boiling point.

It cannot be generally concluded, however, that air heating systems perform better than liquid systems. The collector overall efficiency factor, F_R, is ordinarily lower for air heaters. As a result, X and Y are ordinarily lower and thus the performance of an air system may be equivalent, higher, or lower than that of a liquid system of comparable size.

5.6 DOMESTIC WATER HEATING SYSTEMS

The f-charts presented in Sections 5.3 and 5.4 apply for solar heating systems that supply both space and water heating, although it is assumed that the water heating load is a small fraction, less than about 20%, of the space heating load. A method of estimating the performance of solar heating systems when the heating load is primarily or entirely due to domestic water heating is described here.

The mains water supply temperature, T_m, and the minimum acceptable hot water temperature, T_w, both affect the performance of solar water heating systems. Since both T_m and T_w affect the average system operating temperature level, and thus the collector energy losses, it is reasonable to expect that the dimensionless group X, which has physical significance related to the collector energy losses, can be redefined so as to include the effects of T_m and T_w. If monthly values of X are multiplied by the correction factor in Equation 5.11, the f-chart for liquid-based solar space and water heating systems (Equation 5.5 or Figure 5.1) can be used to estimate monthly values of f for domestic water heating systems.

DOMESTIC WATER HEATING CORRECTION FACTOR = (X_c/X)

$$= \frac{(11.6 + 1.18\, T_w + 3.86\, T_m - 2.32\, \bar{T}_a)}{(100 - \bar{T}_a)} \qquad 5.11$$

The solar water heating systems considered here have a storage capacity of 75 liters of water per square meter of collector area; the average distribution of water usage shown in Figure 4.1 is assumed. These are the conditions upon which the f-chart method is based. The heating load distribution during the day does not have a strong effect upon the performance of solar water heating systems with this size storage capacity. However, the actual distribution of the water heating load may vary greatly from the average distribution shown in Figure 4.1. If most of the daily use typically occurs within a short time interval each day, the load fraction supplied by solar energy may be lower than that estimated using the f-chart in the manner recommended here. In this case, an increase in storage size will increase system performance more than that suggested by the modifications in Section 5.3-2.

Also, the correlation presented here assumes that the water heated above the minimum acceptable temperature, T_w, is no more useful than hot water at a temperature of T_w. At times, the stored water temperature will exceed T_w. It is assumed that the solar energy used to heat water above T_w is wasted and this energy is not considered to be part of the water heating load.

EXAMPLE 5.6 Performance of a Domestic Water Heating System

A solar assisted domestic water heating system is to be installed in a residence located in Madison, Wisconsin. There will be four family members living in the house, and it is anticipated that each person will use 100 liters (25 gallons) of water per day at a temperature of 60 C. The mains supply water temperature in Madison is about 11 C all year. Flat-plate solar collectors having two glass covers and a flat-black absorber surface are being considered. The test results of these collectors are given in Example 2.1. The collectors are to be mounted on the roof of the building facing due south at 58° with respect to horizontal. The capacity of the solar preheat storage tank will be 75 liters of water per square meter of collector area. Estimate the fraction of the water heating load which is supplied by solar energy for collectors areas of 2, 5, and 10 square meters.

The first step is to estimate the monthly heating load. It is assumed that the daily water usage is 100 l/day per person which, for four people, is 400 l/day every day of the year. The average daily water heating load is the product of the daily water usage, the specific heat of water, and the difference between the required hot water temperature, T_w, and the mains supply water temperature, T_m. The average monthly water heating load in column C4 or C5 of worksheet 1 (Table 5.9) is simply the product of the average daily load and the number of days in each month.

From the test results in Example 2.1, it is known that F_RU_L and $F_R(\tau\alpha)_n$ are 3.75 W/C-m^2 and 0.68, respectively. Because of the freezing temperatures in Madison, the system will have an antifreeze solution in the collectors and a heat exchanger between the collectors and the tank. Assuming that both flowrates in the heat exchanger are 0.0139 kg/s-m^2 and the heat exchanger effectiveness is 0.7 at these flowrates, the collector-heat exchanger correction factor, F_R'/F_R, is 0.97 as determined in Example 2.3. Thus, $F_R'U_L$, item C of worksheet 2

TABLE 5.9 WORKSHEET FOR EXAMPLE 5.6

F-CHART WORKSHEET 1

HEATING LOADS

A. $UA = \frac{\text{Design Space Heating Load [W]}}{\text{Design Temperature Difference [°C]}}$ = 0 [W/°C] (See Section 4.2)

B. Water Usage = 400 [liters/day] x 4190 x $(T_w - T_m)$[J/liter] = 82.1×10^6 [J/Day]

Month	C1. Days Per Mo.	C2. Heating Degree Days [°C-day] (from Appendix 2)	C3. Space Heating Load [J/Month] (86400)(A.)(C2.)	C4. Domestic Water Load [J/Month] (See Section 4.3) (B.)(C1.)	C5. Total Load [J/Month] (C3.)+(C4.)
Jan	31			2.55×10^9	
Feb	28			2.30	
Mar	31			2.55	
Apr	30			2.46	
May	31			2.55	
Jun	30			2.46	
Jul	31			2.55	
Aug	31			2.55	
Sep	30			2.46	
Oct	31			2.65	
Nov	30			2.46	
Dec	31			2.55	
Totals	365			30.0×10^9	30.0×10^9

(Table 5.10), and $F_R'(\tau\alpha)_n$, item D, are 3.64 $W/C\text{-}m^2$ and 0.66, respectively.

Monthly values of X/A (column C8) and Y/A (column C9) are determined using Equations 5.3 and 5.4. Monthly average ambient temperatures for Madison can be found in Appendix 2. The difference between 100 C and the monthly ambient temperatures appear in column C7. The values of the average monthly solar radiation on a 58° surface in Madison (in column C10) were calculated in Example 3.1. The ratios of the monthly average to normal incidence transmittance-absorptance products for a two cover collector tilted at 58° in Madison (column C9) were determined in Example 3.3.

For a domestic water heating system, the values of X/A in column C8 must be corrected by the factor

DOMESTIC WATER HEATING CORRECTION FACTOR

$$= \frac{(11.6 + 1.18\ T_w + 3.86\ T_m - 2.32\ \bar{T}_a}{(100 - \bar{T}_a)}$$

where T_w is 60 C and T_m is 11 C in this case. Note that the domestic water heating correction factor has a different value each month because of the variation in the monthly average ambient temperature. The corrected values of X/A appear in column C12 of worksheet 3 (Table 5.11).

Monthly values of X and Y in columns C14 and C15 are determined by multiplying each collector considered by the values of X/A and Y/A in columns C12 and C13. Values of f, the fraction of the monthly water heating load supplied by solar energy, are determined from Equation 5.5 or Figure 5.1 and tabulated in column C16. The total solar energy usefully delivered each month in column C17 is the product of f and the monthly heating load. The annual solar energy usefully delivered and the solar load

TABLE 5.10 WORKSHEET FOR EXAMPLE 5.6

F-CHART WORKSHEET 2

ITEMS MAKING UP X AND Y

C. $F_R U_L (F'_R/F_R)$ = 3.64 [$W/m^2 °C$] (See Sections 2.3 and 2.4)

D. $F_R (\tau\alpha)_n (F'_R/F_R)$ = 0.66 (See Sections 2.3 and 2.4)

	C6. Seconds Per Month	C7. $(100-\overline{T}_a)$[°C] ($\overline{T}_a$ can be found in Appendix 2	C8. X/A[$1/m^2$] $\frac{(C.)(C7.)(C6.)}{(C5.)}$	C9. $(\overline{\tau\alpha})/(\tau\alpha)_n$ (See Sections 3.4 and 3.5)	C10. Daily Radiation on Collector [J/m^2-Day] (See Section 3.2)	C11. Y/A[$1/m^2$] $\frac{(D.)(C9.)(C10.)(C1.)}{(C5.)}$
Jan	2.68×10^6	107	0.410	0.94	13.2×10^6	0.100
Feb	2.42×10^6	106	0.406	0.94	14.8×10^6	0.112
Mar	2.68×10^6	100	0.383	0.93	17.4×10^6	0.130
Apr	2.59×10^6	93	0.356	0.91	15.4×10^6	0.113
May	2.68×10^6	87	0.333	0.90	15.3×10^6	0.111
Jun	2.59×10^6	81	0.310	0.89	16.4×10^6	0.118
Jul	2.68×10^6	79	0.302	0.90	16.7×10^6	0.121
Aug	2.68×10^6	80	0.306	0.90	17.0×10^6	0.126
Sep	2.59×10^6	85	0.326	0.92	18.4×10^6	0.136
Oct	2.68×10^6	90	0.344	0.93	17.0×10^6	0.127
Nov	2.59×10^6	99	0.379	0.94	11.6×10^6	0.088
Dec	2.68×10^6	105	0.402	0.94	12.4×10^6	0.093

TABLE 5.11 WORKSHEET FOR EXAMPLE 5.6

F-CHART WORKSHEET 3

SOLAR HEATING LOAD FRACTION

E. Storage size correction factor (X/X_0) = *See Text*

F. Load heat exchanger correction factor (Y/Y_0) = *1.0* (=1.0 for air systems)

G. Collector air flow rate correction factor (X/X_0) = *1.0* (=1.0 for liquid systems)

	C12.	C13.	Area = 2.0 m²				Area = 5.0 m²				Area = 10.0 m²			
	Corrected X/A	Corrected Y/A	C14.	C15.	C16.	C17.	C14.	C15.	C16.	C17.	C14.	C15.	C16.	C17.
	(C8.)(E.)(G.)	(C11.)(F.)	X	Y	f	(C16.)(C5.)	X	Y	f	(C16.)(C5.)	X	Y	f	(C16.)C5.)
Jan	0.541		1.35	0.25	0.13	0.33×10^9	2.70	0.50	0.30	0.77×10^9	5.41	1.00	0.51	1.30×10^9
Feb	0.531		1.33	0.28	0.15	0.35	2.66	0.56	0.35	0.81	5.31	1.12	0.59	1.36
Mar	0.478	Same as	1.20	0.33	0.19	0.48	2.39	0.65	0.43	1.10	4.78	1.30	0.71	1.81
Apr	0.416		1.04	0.28	0.17	0.42	2.08	0.57	0.39	0.96	4.16	1.13	0.65	1.60
May	0.362	Column C11	0.91	0.28	0.17	0.43	1.81	0.56	0.39	0.99	3.62	1.11	0.66	1.68
Jun	0.309		0.77	0.30	0.19	0.47	1.55	0.59	0.43	1.06	3.09	1.18	0.73	1.80
Jul	0.291	of	0.73	0.30	0.21	0.54	1.45	0.61	0.46	1.17	2.91	1.21	0.77	1.96
Aug	0.300		0.75	0.32	0.21	0.54	1.50	0.63	0.47	1.20	3.00	1.26	0.78	1.99
Sep	0.345	Table 5.10	0.86	0.34	0.22	0.54	1.73	0.68	0.50	1.23	3.45	1.36	0.81	1.99
Oct	0.389		0.97	0.32	0.20	0.51	1.95	0.64	0.45	1.15	3.89	1.27	0.74	1.89
Nov	0.469		1.17	0.22	0.12	0.30	2.35	0.44	0.27	0.66	4.69	0.88	0.47	1.16
Dec	0.522		1.31	0.23	0.12	0.31	2.61	0.47	0.28	0.71	5.22	0.93	0.49	1.25
					Totals	5.2×10^9				11.8×10^9				19.8×10^9

Annual Fractions by Solar (Total, C17.)/(Total, C5.) = 0.17 = 0.39 = 0.66

fraction appear below column C17 for each collector area.

The annual fraction of the total water heating load supplied by solar energy from Table 5.11 is plotted against collector area in Figure 5.8. The general shape of the curve is similar to that of Figure 5.2 for space heating.

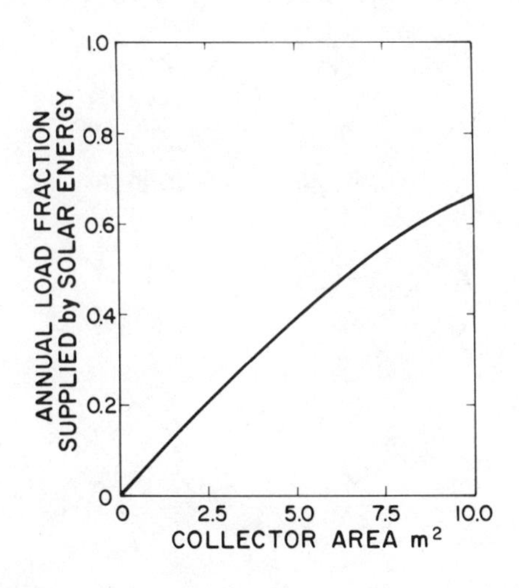

FIGURE 5.8
ANNUAL LOAD FRACTION SUPPLIED BY SOLAR ENERGY

5.7 SUMMARY

Using the f-chart method described in this chapter, the thermal performance of solar space and/or water heating systems can be estimated in a simple manner for any location in which meteorological data are available. The necessary data are given for many North American locations in Appendix 2. The collector parameters required for use with the f-charts are

$F_R(\tau\alpha)_n$ and F_RU_L, which can be determined from standard collector tests.

The results of the thermal performance calculations are best expressed as a plot of F, the fraction of the annual heating load supplied by solar energy, against collector area, such as in Figures 5.2 and 5.8. The information on annual contribution of solar energy as a function of collector area is used in the next chapter to evaluate the economics of solar heating.

CHAPTER 6
SOLAR HEATING ECONOMICS

6.1 INTRODUCTION

Although some people will buy a solar energy heating system to "be independent of the utilities" or "to be the first on their block to own a solar system" most people will consider solar heating only if it is likely to be competitive with the alternatives. The economic choice between the various conventional heating systems can be based entirely on the price of fuel today, if it is assumed that the various choices will all increase in price at roughly the same rate. Usually natural gas (if available) is the least expensive fuel, LP gas and oil are next, and electric resistance heating is the most expensive. Comparing solar with these alternatives on a sound economic basis is the subject of this chapter.

In the simplest terms, a solar energy system will be less expensive than a nonsolar system if the fuel savings are greater than the mortgage payments required to buy the solar equipment. A problem with this comparison is that it is difficult to take into account rising fuel costs. And, there are other items affecting the economics of solar heating which should be taken into account, such as property taxes, income taxes, insurance, and maintenance. A more realistic approach is to use the life cycle cost method which takes into account all future expenses. This method provides a means of comparison of future costs with today's costs, by reducing all costs to the common basis of present worth, that is, what would have to be invested today in order to have the funds available in the future to meet all of the anticipated expenses.

In a life cycle cost analysis, the required cash flow (net payment) for heating is calculated for each year. The life cycle cost is found by discounting each annual cash flow to its present value and finding the sum of these discounted cash flows. When the present values of all future costs have been determined for each of the alternative systems under consideration, including solar and nonsolar options, the system that yields the lowest life cycle cost is selected as the most cost effective. The reason that cash flows must be discounted lies in the "time value of money." A fuel bill or other expense that is anticipated to be $1000 in 10 years is equivalent to an obligation of $463 today at a "market discount rate" of 8%. In other words, an investment of $463 today at an 8% interest rate will have a value of $1000 in 10 years.

Although life cycle costing is accepted by most economists as the soundest approach for making an economic decision (e.g., see Ruegg (1975)), it has one major drawback; we must be able to predict costs into the future. Future costs of fuel and other expenses and the market discount rate must be estimated. Given the uncertainty in predicting future costs, one alternative is to make a set of pessimistic and another set of optimistic assumptions and determine the most cost effective system for both cases. The choice between the designs is then made on "intuition" by the homeowner or the businessman.

The computation involved in a life cycle cost analysis is significant. A large business has the trained personnel and can easily afford the effort since they are usually considering large installations. Although the typical homeowner or small businessman may not be able to make these calculations, he can understand the factors involved and he can appreciate the "bottom line." His building contractor, architect, or heating and ventilating engineer will have to do the calculations and explain the consequences of various decisions. In order to ease the computational burden involved in estimating the cost of solar heating, worksheets are provided, which for most systems permit the life cycle cost calculations to be no more time consuming than a single year's calculation.

6.2 GENERAL PROCEDURE

In order to remove some of the complexity in presenting this material, we will first look at the yearly cost of heating. This yearly cost includes all necessary payments to heat the building, less any tax savings. Tax savings resulting from income tax deductions are an important consideration and depend upon who owns the system. For a homeowner with a nonsolar system, his income tax deductions, for the purpose of this calculation, are limited to the interest and property tax that can be ascribed to the heating system. For the owner of an income-producing property, all expenses, including fuel, are deductable.

The yearly expenses to heat a building include the annual payment for interest and principle to buy the equipment, the fuel expense, insurance, maintaince costs, and possibly property taxes. This yearly expense is reduced by tax savings, which for a homeowner, consists of income tax deductions for interest payment and property taxes. For a businessman with an income-producing property, his income tax deductions will include all income-producing expenses plus a depreciation allowance. Both the homeowner and the businessman may be able to take advantage of state and/or federal tax credits specifically designed to encourage solar energy utilization.

In equation form, the annual costs for both solar and nonsolar systems can be expressed as:

$$\begin{gathered}\text{Yearly cost for heating} = \text{Mortage payment} + \text{Fuel expense} + \text{Maintenance and insurance} \\ + \text{Property tax} - \text{Tax savings} \qquad (6.1)\end{gathered}$$

where the income tax savings for a residential building are given by

$$\text{Tax savings} = \text{Tax rate} \times \left[\text{Interest payment} + \text{Property tax}\right] \qquad (6.2)$$

and the income tax savings for a commercial installation are

$$\begin{gathered}\text{Tax savings} = \text{Tax rate} \times \Big[\text{Interest payment} + \text{Property tax} + \text{Fuel expense} \\ + \text{Maint.} + \text{Ins.} + \text{Deprec.}\Big] \qquad (6.3)\end{gathered}$$

The most general economic analysis will evaluate each heating alternative. An economic decision can then be made by comparing the life cycle costs of these alternatives. With only a small loss in generality, the concept of solar savings can be used to simplify some of the calculations. Solar savings are the difference between the life cycle cost of a conventional heating system and a solar heating system. (Savings can be negative; they are then losses.) The savings concept is useful because it is not necessary to evaluate costs that are common to both the solar and the nonsolar system. For example, the auxiliary furnace and much of the ductwork and plumbing in a solar system are often the same as would be installed in a nonsolar system. With the savings concept, it is only necessary to estimate the incremental cost of installing a solar system. If the furnaces or other equipment in the two systems are different, the difference in their costs can be included as an extra cost (or savings) of installing a solar system. In equation form, solar savings are

$$\begin{array}{c}\text{Solar}\\\text{savings}\end{array} = \begin{array}{c}\text{Fuel}\\\text{savings}\end{array} - \begin{array}{c}\text{Extra mort.}\\\text{payment}\end{array} - \begin{array}{c}\text{Extra ins. \&}\\\text{maintenance}\end{array} - \begin{array}{c}\text{Extra}\\\text{prop. tax}\end{array} + \begin{array}{c}\text{Tax}\\\text{savings}\end{array} \qquad 6.4$$

The income tax savings for a residence are

$$\begin{array}{c}\text{Tax}\\\text{savings}\end{array} = \begin{array}{c}\text{Tax}\\\text{rate}\end{array} \times \left[\begin{array}{c}\text{Extra}\\\text{interest}\end{array} + \begin{array}{c}\text{Extra}\\\text{property tax}\end{array}\right] \qquad 6.5$$

and the income tax savings for a business are

$$\begin{array}{c}\text{Tax}\\\text{savings}\end{array} = \begin{array}{c}\text{Tax}\\\text{rate}\end{array} \times \left[\begin{array}{c}\text{Extra}\\\text{interest}\end{array} + \begin{array}{c}\text{Extra}\\\text{maintenance}\end{array} + \begin{array}{c}\text{Extra}\\\text{insurance}\end{array} + \begin{array}{c}\text{Extra}\\\text{property tax}\end{array} + \begin{array}{c}\text{Extra}\\\text{depreciation}\end{array} - \begin{array}{c}\text{Fuel}\\\text{saved}\end{array}\right] \qquad 6.6$$

The reason that fuel saved is a negative tax deduction is that a business already deducts fuel expenses so that the cost of fuel saved is taxable income.

These equations must be evaluated for each year of the period of the economic analysis, and each resulting yearly savings must be discounted back to the present time and summed. This calculation is illustrated by Table 6.1. Here we show the terms in Equation 6.4 for the first few years and the last year. Column 1 is the fuel savings, 2 is extra mortgage payment, 3 is extra insurance and maintenance, 4 is extra property tax, 5 is tax savings, 6 is the solar savings, and 7 is the present worth of each of the items in column 6, that is, the amount of money which would have to be invested today to have the amount in column 6 available in that year.

TABLE 6.1
EXAMPLE OF A SOLAR SAVINGS
AND PRESENT WORTH CALCULATION

	1	2	3	4	5	6	7
YEAR	FUEL SAV.	EXTRA MORT. PAY.	EXTRA INS. & MAINT.	EXTRA PROP. TAX	TAX SAV.	SOLAR SAV.	PRES. WORTH OF 6
1	457	592	60	80	260	-615	-614*
2	502	592	64	85	258	19	16
3	553	592	67	90	256	60	48
4	608	592	71	95	253	103	76
.	.	.	.	.	.	.	.
.	.	.	.	.	.	.	.
.	.	.	.	.	.	.	.
20	3074	592	192	258	141	2173	466
						LIFE CYCLE SOLAR SAVINGS	$3801

*The first year includes a downpayment of $600 at the beginning of the year and is not discounted.

In this example, we have assumed a market discount rate of 8%. Thus, for example, column 6 of the table indicates that $103 will be the solar savings in the fourth year; the present worth of these savings is $103/(1.08)^4=\$76$.

Each of the individual entries in this table is an estimate of what that particular item is expected to be in a future year. Somehow, estimates must be made of these anticipated costs and gains; only by doing so can the comparisons of future costs and present investments be made. Two situations may occur.

First, costs may be anticipated to change in a regular manner through the period of the analysis. The most common assumption would be that each type of cost represented as a term in Equations 6.4 to 6.6 inflates or deflates at a fixed percentage per year. The calculations for this situation are simplified, and it is not necessary to make up a table such as Table 6.1. The sum of the present worth of the items in each column (e.g., fuel savings) can be calculated by a convenient equation and these present worth items can then be used to determine the life cycle solar savings. This method is outlined in sections 6.3 to 6.7.

Second, if costs are expected to change in an irregular manner, or if incremental costs are anticipated during the period of the analysis, then a table such as Table 6.1 may have to be used. However, if the anticipated costs are not too irregular, then simplified methods can be used. Examples of this would be the anticipation of major maintenance expenses at regular intervals during the period, or an expected change in the rate at which fuel prices increase. A method for handling these situations is outlined in Section 6.9.

6.3 REGULARLY VARYING COSTS

If costs are assumed to inflate (or deflate) at a fixed percentage each year, the life cycle savings can be obtained by calculations that are only slightly more complicated than the calculations for a single year. To perform this calculation we have tabulated an inflation-discount function to inflate at a rate i per period, discount at a rate d per period, and sum each economic term for N periods. (Note that the period N is usually a year, and d and i are then yearly rates.) This function is defined as:

$$\text{FUNCTION}(N,i,d) = \frac{1}{d-i}\left[1 - \left(\frac{1+i}{1+d}\right)^N\right] \qquad i \neq d \qquad 6.7$$

$$= N/(1+i) \qquad i = d$$

and is tabulated in increments of 5 for N from 5 to 30 in Tables 6.2A through 6.2F.

If the inflation-discount function of Equation 6.7 is multiplied by the value of the first payment of any expense that is expected to inflate at a rate i, then the result is the sum of N such payments, discounted to the present time with a discount rate of d. As an example, we can obtain the 20-year life cycle fuel expense with fuel inflating at 10% a year and with a market discount rate of 8% by multiplying the first year's fuel expense by FUNCTION(20,0.10,0.08) = 22.169 (from Table 6.2D). Life cycle maintenance costs, insurance costs and property taxes can also be evaluated in this manner as long as they are assumed to inflate at fixed rates.

This equation assumes that payments are made at the end of each time period (e.g. December 31) and the resulting present worth is as of the beginning of the first time period (e.g. January 1 of the first year). Note that the first year's expense is evaluated at the end of the first time period. These ideas are illustrated on Figure 6.1. Some authors (Rueggs, 1975) have defined this function with an additional multiplier of (1+i). There is no fundamental problem with either definition as long as the assumptions are clearly defined.

These tables have other uses. They can be used to find the uniform annual payment of a loan by dividing the loan principle by FUNCTION(N,0,m), where m is the mortgage interest rate and N is the number of years. This is found in Table 6.2 using a market discount rate of m and an annual inflation rate of zero. For example, the annual payment on a $1000, 20-year, 8% mortgage is $1000/9.818 or $101.85. We can then find the present value of 20 such payments with a 6% discount rate by multiplying $101.85 by FUNCTION(20,0,0.06) or 11.470 to give $1168. In other words, the true cost of this $1000 loan is $168. This $168 will be further reduced if we consider tax deductions for interest.

These tables can also be used to find the present value of all interest paid on a loan. The calculation

TABLE 6.2A INFLATION-DISCOUNT FUNCTION FOR N = 5

d, MARKET DISCOUNT RATE (%)	i, ANNUAL INFLATION RATE (%)												
	0	1	2	3	4	5	6	7	8	9	10	11	12
0	5.000	5.101	5.204	5.309	5.416	5.526	5.637	5.751	5.867	5.985	6.105	6.228	6.353
1	4.853	4.950	5.049	5.150	5.253	5.358	5.466	5.575	5.686	5.799	5.915	6.033	6.153
2	4.713	4.807	4.902	4.999	5.098	5.199	5.302	5.407	5.514	5.623	5.734	5.847	5.962
3	4.580	4.669	4.761	4.854	4.950	5.047	5.146	5.246	5.349	5.454	5.561	5.669	5.780
4	4.452	4.538	4.626	4.716	4.808	4.901	4.996	5.093	5.192	5.293	5.395	5.500	5.606
5	4.329	4.413	4.497	4.584	4.672	4.762	4.853	4.947	5.042	5.139	5.238	5.338	5.441
6	4.212	4.292	4.374	4.457	4.542	4.629	4.717	4.807	4.898	4.992	5.087	5.183	5.282
7	4.100	4.177	4.256	4.336	4.418	4.501	4.586	4.673	4.761	4.851	4.942	5.036	5.131
8	3.993	4.067	4.143	4.220	4.299	4.379	4.461	4.545	4.630	4.716	4.804	4.894	4.986
9	3.890	3.961	4.035	4.109	4.185	4.263	4.342	4.422	4.504	4.587	4.672	4.759	4.847
10	3.791	3.860	3.931	4.003	4.076	4.151	4.227	4.304	4.383	4.464	4.545	4.629	4.714
11	3.696	3.763	3.831	3.900	3.971	4.043	4.117	4.191	4.268	4.345	4.424	4.505	4.586
12	3.605	3.669	3.735	3.802	3.870	3.940	4.011	4.083	4.157	4.231	4.308	4.385	4.464
13	3.517	3.580	3.643	3.708	3.774	3.841	3.909	3.979	4.050	4.122	4.196	4.271	4.347
14	3.433	3.493	3.555	3.617	3.681	3.746	3.812	3.879	3.948	4.018	4.089	4.161	4.235
15	3.352	3.410	3.470	3.530	3.592	3.655	3.719	3.784	3.850	3.917	3.986	4.056	4.127
16	3.274	3.331	3.388	3.447	3.506	3.567	3.629	3.691	3.755	3.821	3.887	3.954	4.023
17	3.199	3.254	3.309	3.366	3.424	3.482	3.542	3.603	3.665	3.728	3.792	3.857	3.924
18	3.127	3.180	3.234	3.288	3.344	3.401	3.459	3.518	3.577	3.638	3.700	3.764	3.828
19	3.058	3.109	3.161	3.214	3.268	3.323	3.379	3.436	3.493	3.552	3.612	3.673	3.736
20	2.991	3.040	3.091	3.142	3.194	3.247	3.301	3.357	3.413	3.470	3.528	3.587	3.647

TABLE 6.2B INFLATION-DISCOUNT FUNCTION FOR N = 10

d, MARKET DISCOUNT RATE (%)	i, ANNUAL INFLATION RATE (%) 0	1	2	3	4	5	6	7	8	9	10	11	12
0	10.000	10.462	10.950	11.464	12.006	12.578	13.181	13.816	14.487	15.193	15.937	16.722	17.549
1	9.471	9.901	10.354	10.831	11.335	11.865	12.425	13.014	13.635	14.289	14.979	15.705	16.470
2	8.983	9.383	9.804	10.248	10.716	11.209	11.728	12.275	12.851	13.458	14.097	14.770	15.479
3	8.530	8.903	9.295	9.709	10.144	10.603	11.085	11.594	12.129	12.692	13.286	13.910	14.567
4	8.111	8.459	8.825	9.210	9.615	10.042	10.492	10.965	11.462	11.986	12.537	13.117	13.727
5	7.722	8.046	8.388	8.748	9.126	9.524	9.942	10.383	10.846	11.334	11.847	12.386	12.953
6	7.360	7.664	7.983	8.319	8.672	9.043	9.434	9.845	10.277	10.731	11.208	11.710	12.238
7	7.024	7.308	7.607	7.921	8.251	8.598	8.962	9.346	9.749	10.172	10.618	11.085	11.577
8	6.710	6.976	7.256	7.550	7.859	8.184	8.525	8.883	9.259	9.655	10.070	10.507	10.965
9	6.418	6.667	6.930	7.205	7.495	7.798	8.118	8.453	8.805	9.174	9.562	9.970	10.398
10	6.145	6.379	6.625	6.884	7.155	7.440	7.739	8.053	8.382	8.728	9.091	9.472	9.872
11	5.889	6.110	6.341	6.584	6.838	7.105	7.386	7.680	7.989	8.313	8.652	9.009	9.383
12	5.650	5.858	6.075	6.303	6.543	6.793	7.057	7.333	7.622	7.926	8.244	8.578	8.929
13	5.426	5.622	5.826	6.041	6.266	6.502	6.749	7.008	7.280	7.565	7.864	8.177	8.505
14	5.216	5.400	5.593	5.795	6.007	6.229	6.462	6.705	6.961	7.228	7.509	7.803	8.111
15	5.019	5.193	5.374	5.565	5.765	5.974	6.193	6.422	6.662	6.914	7.177	7.453	7.743
16	4.833	4.997	5.169	5.349	5.537	5.734	5.940	6.156	6.383	6.619	6.867	7.127	7.399
17	4.659	4.814	4.976	5.146	5.323	5.509	5.704	5.908	6.121	6.344	6.577	6.822	7.077
18	4.494	4.641	4.794	4.955	5.123	5.298	5.482	5.674	5.875	6.085	6.305	6.536	6.776
19	4.339	4.478	4.623	4.775	4.934	5.100	5.273	5.455	5.644	5.843	6.050	6.267	6.494
20	4.192	4.324	4.462	4.606	4.756	4.913	5.077	5.248	5.428	5.615	5.811	6.016	6.230

TABLE 6.2C INFLATION-DISCOUNT FUNCTION FOR N = 15

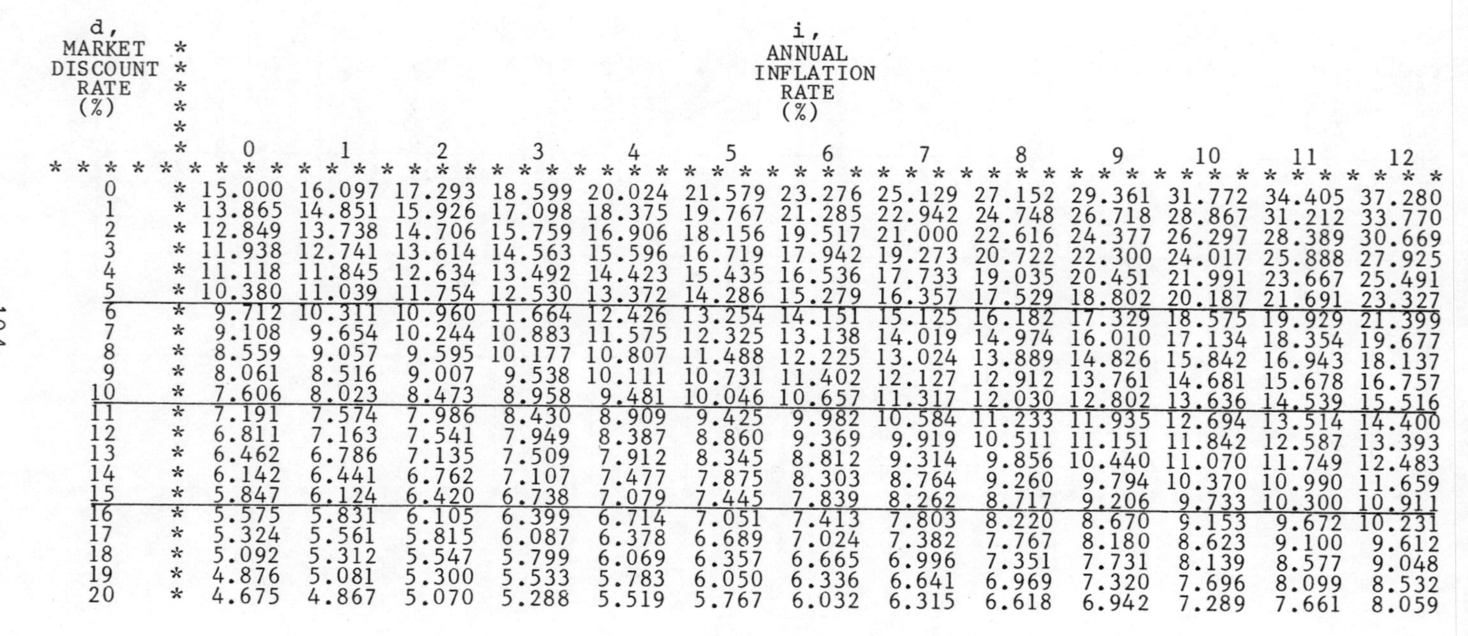

d, MARKET DISCOUNT RATE (%)	i, ANNUAL INFLATION RATE (%) 0	1	2	3	4	5	6	7	8	9	10	11	12
0	15.000	16.097	17.293	18.599	20.024	21.579	23.276	25.129	27.152	29.361	31.772	34.405	37.280
1	13.865	14.851	15.926	17.098	18.375	19.767	21.285	22.942	24.748	26.718	28.867	31.212	33.770
2	12.849	13.738	14.706	15.759	16.906	18.156	19.517	21.000	22.616	24.377	26.297	28.389	30.669
3	11.938	12.741	13.614	14.563	15.596	16.719	17.942	19.273	20.722	22.300	24.017	25.888	27.925
4	11.118	11.845	12.634	13.492	14.423	15.435	16.536	17.733	19.035	20.451	21.991	23.667	25.491
5	10.380	11.039	11.754	12.530	13.372	14.286	15.279	16.357	17.529	18.802	20.187	21.691	23.327
6	9.712	10.311	10.960	11.664	12.426	13.254	14.151	15.125	16.182	17.329	18.575	19.929	21.399
7	9.108	9.654	10.244	10.883	11.575	12.325	13.138	14.019	14.974	16.010	17.134	18.354	19.677
8	8.559	9.057	9.595	10.177	10.807	11.488	12.225	13.024	13.889	14.826	15.842	16.943	18.137
9	8.061	8.516	9.007	9.538	10.111	10.731	11.402	12.127	12.912	13.761	14.681	15.678	16.757
10	7.606	8.023	8.473	8.958	9.481	10.046	10.657	11.317	12.030	12.802	13.636	14.539	15.516
11	7.191	7.574	7.986	8.430	8.909	9.425	9.982	10.584	11.233	11.935	12.694	13.514	14.400
12	6.811	7.163	7.541	7.949	8.387	8.860	9.369	9.919	10.511	11.151	11.842	12.587	13.393
13	6.462	6.786	7.135	7.509	7.912	8.345	8.812	9.314	9.856	10.440	11.070	11.749	12.483
14	6.142	6.441	6.762	7.107	7.477	7.875	8.303	8.764	9.260	9.794	10.370	10.990	11.659
15	5.847	6.124	6.420	6.738	7.079	7.445	7.839	8.262	8.717	9.206	9.733	10.300	10.911
16	5.575	5.831	6.105	6.399	6.714	7.051	7.413	7.803	8.220	8.670	9.153	9.672	10.231
17	5.324	5.561	5.815	6.087	6.378	6.689	7.024	7.382	7.767	8.180	8.623	9.100	9.612
18	5.092	5.312	5.547	5.799	6.069	6.357	6.665	6.996	7.351	7.731	8.139	8.577	9.048
19	4.876	5.081	5.300	5.533	5.783	6.050	6.336	6.641	6.969	7.320	7.696	8.099	8.532
20	4.675	4.867	5.070	5.288	5.519	5.767	6.032	6.315	6.618	6.942	7.289	7.661	8.059

TABLE 6.2D INFLATION-DISCOUNT FUNCTION FOR N = 20

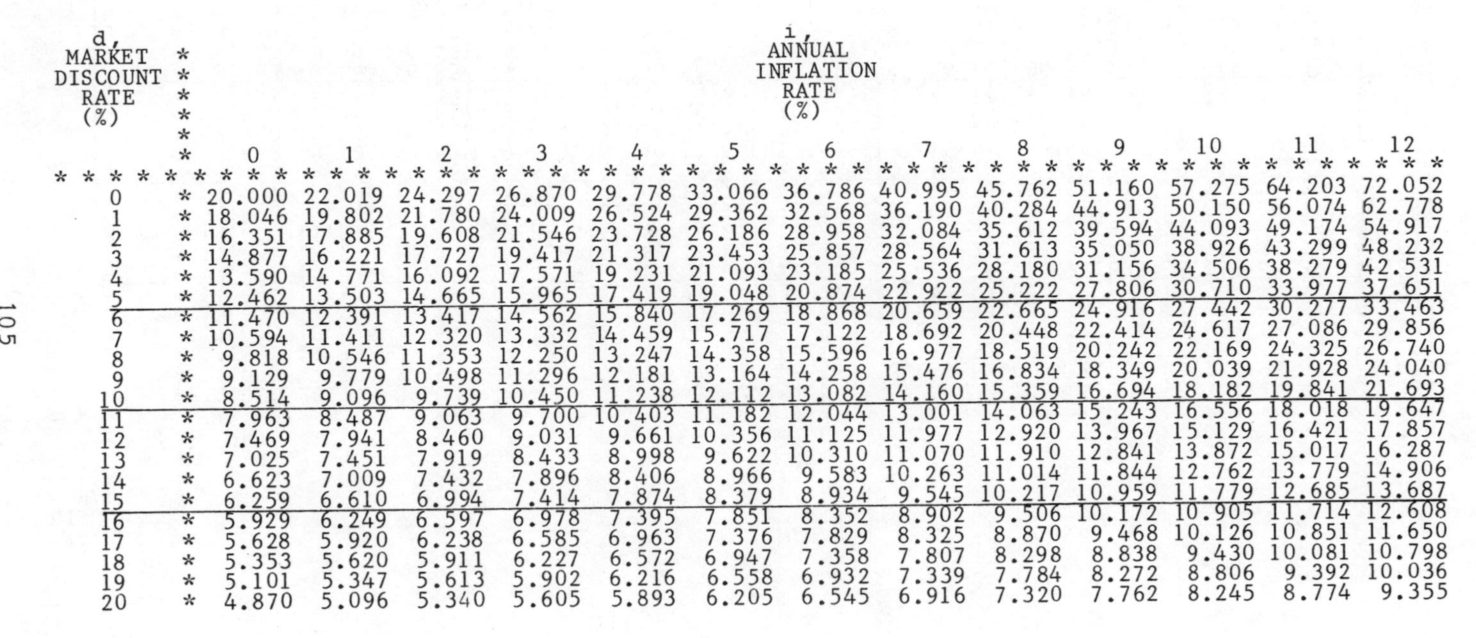

d, MARKET DISCOUNT RATE (%)	i, ANNUAL INFLATION RATE (%) 0	1	2	3	4	5	6	7	8	9	10	11	12
0	20.000	22.019	24.297	26.870	29.778	33.066	36.786	40.995	45.762	51.160	57.275	64.203	72.052
1	18.046	19.802	21.780	24.009	26.524	29.362	32.568	36.190	40.284	44.913	50.150	56.074	62.778
2	16.351	17.885	19.608	21.546	23.728	26.186	28.958	32.084	35.612	39.594	44.093	49.174	54.917
3	14.877	16.221	17.727	19.417	21.317	23.453	25.857	28.564	31.613	35.050	38.926	43.299	48.232
4	13.590	14.771	16.092	17.571	19.231	21.093	23.185	25.536	28.180	31.156	34.506	38.279	42.531
5	12.462	13.503	14.665	15.965	17.419	19.048	20.874	22.922	25.222	27.806	30.710	33.977	37.651
6	11.470	12.391	13.417	14.562	15.840	17.269	18.868	20.659	22.665	24.916	27.442	30.277	33.463
7	10.594	11.411	12.320	13.332	14.459	15.717	17.122	18.692	20.448	22.414	24.617	27.086	29.856
8	9.818	10.546	11.353	12.250	13.247	14.358	15.596	16.977	18.519	20.242	22.169	24.325	26.740
9	9.129	9.779	10.498	11.296	12.181	13.164	14.258	15.476	16.834	18.349	20.039	21.928	24.040
10	8.514	9.096	9.739	10.450	11.238	12.112	13.082	14.160	15.359	16.694	18.182	19.841	21.693
11	7.963	8.487	9.063	9.700	10.403	11.182	12.044	13.001	14.063	15.243	16.556	18.018	19.647
12	7.469	7.941	8.460	9.031	9.661	10.356	11.125	11.977	12.920	13.967	15.129	16.421	17.857
13	7.025	7.451	7.919	8.433	8.998	9.622	10.310	11.070	11.910	12.841	13.872	15.017	16.287
14	6.623	7.009	7.432	7.896	8.406	8.966	9.583	10.263	11.014	11.844	12.762	13.779	14.906
15	6.259	6.610	6.994	7.414	7.874	8.379	8.934	9.545	10.217	10.959	11.779	12.685	13.687
16	5.929	6.249	6.597	6.978	7.395	7.851	8.352	8.902	9.506	10.172	10.905	11.714	12.608
17	5.628	5.920	6.238	6.585	6.963	7.376	7.829	8.325	8.870	9.468	10.126	10.851	11.650
18	5.353	5.620	5.911	6.227	6.572	6.947	7.358	7.807	8.298	8.838	9.430	10.081	10.798
19	5.101	5.347	5.613	5.902	6.216	6.558	6.932	7.339	7.784	8.272	8.806	9.392	10.036
20	4.870	5.096	5.340	5.605	5.893	6.205	6.545	6.916	7.320	7.762	8.245	8.774	9.355

TABLE 6.2E INFLATION-DISCOUNT FUNCTION FOR N = 25

d, MARKET DISCOUNT RATE (%)	i, ANNUAL INFLATION RATE (%) 0	1	2	3	4	5	6	7	8	9	10	11	12
0	25.000	28.243	32.030	36.459	41.646	47.727	54.864	63.249	73.106	84.701	98.347	114.413	133.334
1	22.023	24.752	27.929	31.633	35.958	41.014	46.933	53.869	62.003	71.550	82.762	95.935	111.419
2	19.523	21.832	24.510	27.622	31.245	35.470	40.401	46.164	52.906	60.800	70.051	80.897	93.621
3	17.413	19.375	21.644	24.272	27.322	30.867	34.994	39.804	45.417	51.974	59.639	68.606	79.104
4	15.622	17.298	19.229	21.459	24.038	27.028	30.498	34.531	39.224	44.693	51.071	58.516	67.213
5	14.094	15.532	17.184	19.085	21.277	23.810	26.740	30.137	34.079	38.660	43.990	50.197	57.431
6	12.783	14.024	15.444	17.072	18.943	21.098	23.585	26.458	29.784	33.639	38.112	43.308	49.350
7	11.654	12.729	13.954	15.356	16.961	18.803	20.923	23.364	26.183	29.440	33.210	37.578	42.645
8	10.675	11.611	12.674	13.885	15.269	16.851	18.666	20.750	23.148	25.912	29.103	32.791	37.058
9	9.823	10.641	11.568	12.620	13.817	15.182	16.743	18.530	20.580	22.936	25.648	28.774	32.382
10	9.077	9.796	10.607	11.525	12.566	13.749	15.097	16.636	18.396	20.412	22.727	25.388	28.452
11	8.422	9.056	9.769	10.574	11.482	12.512	13.682	15.012	16.530	18.264	20.248	22.523	25.134
12	7.843	8.405	9.035	9.743	10.540	11.440	12.459	13.615	14.929	16.425	18.133	20.086	22.321
13	7.330	7.830	8.388	9.014	9.716	10.506	11.398	12.406	13.548	14.846	16.322	18.005	19.926
14	6.873	7.320	7.817	8.372	8.993	9.689	10.473	11.356	12.353	13.483	14.764	16.220	17.878
15	6.464	6.865	7.309	7.803	8.355	8.971	9.662	10.439	11.314	12.301	13.417	14.683	16.119
16	6.097	6.457	6.856	7.298	7.790	8.338	8.950	9.636	10.406	11.272	12.249	13.353	14.602
17	5.766	6.092	6.451	6.848	7.288	7.776	8.321	8.929	9.609	10.372	11.230	12.197	13.288
18	5.467	5.762	6.086	6.444	6.839	7.277	7.763	8.304	8.907	9.582	10.339	11.189	12.146
19	5.195	5.463	5.758	6.081	6.437	6.830	7.266	7.749	8.286	8.886	9.556	10.306	11.148
20	4.948	5.192	5.460	5.753	6.075	6.430	6.822	7.255	7.735	8.269	8.864	9.529	10.272

TABLE 6.2F INFLATION-DISCOUNT FUNCTION FOR N = 30

d, MARKET DISCOUNT RATE (%)	i, ANNUAL INFLATION RATE (%) 0	1	2	3	4	5	6	7	8	9	10	11	12
0	30.000	34.785	40.568	47.575	56.085	66.439	79.058	94.461	113.283	136.307	164.494	199.021	241.333
1	25.808	29.703	34.389	40.042	46.878	55.164	65.225	77.462	92.367	110.545	132.735	159.843	192.981
2	22.396	25.589	29.412	34.002	39.529	46.201	54.270	64.050	75.922	90.353	107.916	129.313	155.400
3	19.600	22.235	25.374	29.126	33.624	39.029	45.541	53.404	62.914	74.435	88.413	105.392	126.034
4	17.292	19.481	22.076	25.163	28.846	33.254	38.541	44.900	52.563	61.813	73.000	86.545	102.965
5	15.372	17.203	19.363	21.919	24.955	28.571	32.891	38.065	44.276	51.746	60.748	71.613	84.744
6	13.765	15.307	17.116	19.246	21.765	24.751	28.302	32.537	37.601	43.668	50.953	59.716	70.272
7	12.409	13.716	15.241	17.028	19.131	21.612	24.549	28.037	32.190	37.147	43.076	50.182	58.715
8	11.258	12.372	13.667	15.176	16.942	19.017	21.461	24.351	27.778	31.851	36.704	42.499	49.433
9	10.274	11.230	12.335	13.618	15.111	16.856	18.904	21.313	24.157	27.523	31.518	36.271	41.937
10	9.427	10.253	11.202	12.299	13.569	15.046	16.771	18.792	21.166	23.965	27.273	31.192	35.848
11	8.694	9.411	10.232	11.175	12.262	13.520	14.982	16.687	18.681	21.022	23.776	27.027	30.873
12	8.055	8.682	9.395	10.211	11.147	12.225	13.472	14.918	16.603	18.572	20.879	23.590	26.786
13	7.496	8.046	8.670	9.379	10.190	11.119	12.188	13.423	14.855	16.520	18.464	20.738	23.407
14	7.003	7.489	8.037	8.658	9.363	10.169	11.091	12.151	13.375	14.792	16.438	18.356	20.599
15	6.566	6.997	7.482	8.028	8.646	9.347	10.147	11.063	12.115	13.327	14.729	16.356	18.250
16	6.177	6.562	6.992	7.475	8.019	8.633	9.331	10.126	11.035	12.078	13.279	14.667	16.275
17	5.829	6.174	6.558	6.987	7.468	8.009	8.621	9.315	10.104	11.007	12.041	13.231	14.605
18	5.517	5.827	6.171	6.554	6.981	7.460	7.999	8.608	9.298	10.083	10.979	12.005	13.184
19	5.235	5.515	5.825	6.168	6.550	6.976	7.453	7.990	8.596	9.282	10.061	10.951	11.968
20	4.979	5.233	5.513	5.822	6.165	6.545	6.970	7.446	7.980	8.583	9.265	10.040	10.922

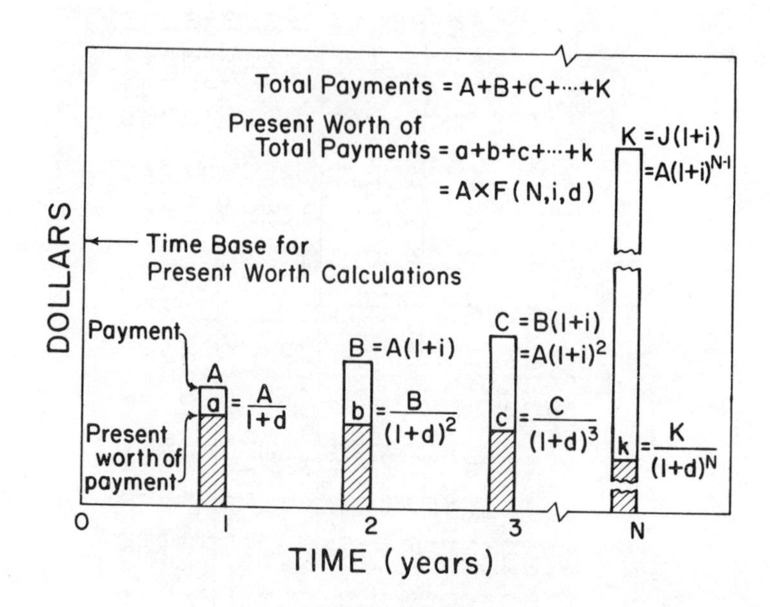

FIGURE 6.1
PRESENT WORTH OF A UNIFORM
SERIES OF INFLATING PAYMENTS

requires a number of values from Table 6.2 and will be illustrated in Example 6.1. When these calculations are performed for the previous $1000 loan, we find that the present value of all the interest is $586. If the borrower's average effective tax bracket is 40%, then the present value of his tax savings is 0.4x586=$235. An alternative way of looking at the tax savings is to compare it with $168, the true cost of the loan. We find that under these economic and tax conditions, borrowing $1000 actually results in a "profit" of 235-168 or $67.

As will be demonstrated in the examples, Table 6.2 can be used to find the present value of the straight-line depreciation allowance on commercial property. If sum-of-digits or double-declining balance depreciation methods are used, Tables 6.3A or 6.3B must be used.

Worksheets are provided as a convenience in doing the economic calculations. The use of these worksheets is illustrated in the following examples, and blank worksheets are included in Appendix 5. Worksheet 4 includes, in items H through Z, listing of the

TABLE 3.3A SUM-OF-DIGITS DEPRECIATION FACTOR

MARKET DISCOUNT RATE %	YEARS OF DEPRECIATION					
	5	10	15	20	25	30
0	5.0000	10.0000	15.0000	20.0000	25.0000	30.0000
1	4.8856	9.6126	14.1868	18.6138	22.8988	27.0470
2	4.7757	9.2492	13.4421	17.3741	21.0636	24.5276
3	4.6699	8.9079	12.7586	16.2620	19.4535	22.3646
4	4.5681	8.5868	12.1300	15.2611	18.0345	20.4967
5	4.4702	8.2846	11.5509	14.3577	16.7785	18.8743
6	4.3758	7.9997	11.0161	13.5398	15.6624	17.4572
7	4.2848	7.7310	10.5216	12.7973	14.6664	16.2129
8	4.1970	7.4771	10.0633	12.1213	13.7743	15.1147
9	4.1124	7.2371	9.6379	11.5042	12.9722	14.1407
10	4.0307	7.0099	9.2424	10.9395	12.2484	13.2730
11	3.9518	6.7947	8.8740	10.4214	11.5932	12.4963
12	3.8756	6.5906	8.5303	9.9449	10.9980	11.7983
13	3.8020	6.3969	8.2093	9.5057	10.4556	11.1684
14	3.7308	6.2128	7.9088	9.0999	9.9599	10.5978
15	3.6619	6.0379	7.6272	8.7242	9.5056	10.0791
16	3.5952	5.8713	7.3629	8.3757	9.0879	9.6060
17	3.5307	5.7127	7.1146	8.0517	8.7031	9.1729
18	3.4682	5.5615	6.8808	7.7499	8.3475	8.7753
19	3.4077	5.4173	6.6606	7.4682	8.0182	8.4093
20	3.3490	5.2796	6.4528	7.2050	7.7125	8.0713

TABLE 3.3B DOUBLE-DECLINING BALANCE DEPRECIATION FACTOR

MARKET DISCOUNT RATE %	YEARS OF DEPRECIATION					
	5	10	15	20	25	30
0	5.0000	10.0000	15.0000	20.0000	25.0000	30.0000
1	4.8871	9.5701	14.0589	18.3630	22.4916	26.4534
2	4.7787	9.1710	13.2134	16.9394	20.3790	23.5593
3	4.6746	8.7999	12.4516	15.6953	18.5868	21.1738
4	4.5745	8.4542	11.7631	14.6028	17.0554	19.1878
5	4.4783	8.1318	11.1391	13.6388	15.7377	17.5183
6	4.3857	7.8307	10.5719	12.7843	14.5962	16.1017
7	4.2966	7.5489	10.0548	12.0234	13.6007	14.8889
8	4.2108	7.2850	9.5823	11.3430	12.7270	13.8416
9	4.1280	7.0373	9.1492	10.7318	11.9556	12.9299
10	4.0483	6.8047	8.7513	10.1807	11.2705	12.1302
11	3.9713	6.5858	8.3848	9.6818	10.6588	11.4237
12	3.8971	6.3796	8.0465	9.2284	10.1097	10.7955
13	3.8253	6.1852	7.7333	8.8149	9.6145	10.2335
14	3.7561	6.0016	7.4429	8.4365	9.1657	9.7278
15	3.6891	5.8280	7.1729	8.0891	8.7572	9.2704
16	3.6243	5.6637	6.9214	7.7692	8.3840	8.8547
17	3.5617	5.5080	6.6866	7.4737	8.0417	8.4752
18	3.5010	5.3603	6.4671	7.1999	7.7267	8.1274
19	3.4423	5.2201	6.2614	6.9457	7.4357	7.8074
20	3.3854	5.0867	6.0683	6.7089	7.1661	7.5120

various parameters which affect the economics of the solar heating system. Items AA through FF are values of FUNCTION(N,i,d), obtained from Table 6.2 or Equation 6.7. AA is for fuel costs, while BB through FF all relate to the investment and are used in calculating life cycle expenses as a fraction of the initial investment. Items GG through PP are a convenient way of doing the arithmetic needed in Equations 6.4 through 6.7. Then, Worksheet 5 utilizes information from Worksheet 4, AA and either OO or PP, in the calculation of the solar savings for systems of various collector areas.

EXAMPLE 6.1 Solar Economics for a Residence

We will assume that the building of Example 4.1 is a residence. The reason for using a residence is that the income tax deductions for commercial property are more complicated than for a residence. This additional tax calculation is the only difference between the life cycle cost of a small office and a residence and will be illustrated in example 6.2. A completed version of worksheet 4 is shown in Table 6.4. Only those entries that are not self explanatory will be discussed. (The only entries that use S.I. units are line K, which could be changed to $ per square foot, and lines M and N, which could be changed to $/million BTU.)

Lines K and L represent a simple way of looking at the solar investment. For many installations, the costs can be placed into two categories; a cost that is dependent on collector area and a fixed cost. The area-dependent cost includes the cost of the collector, labor for installation, possible credits for savings on the roof, expenses for roof support modifications, part of the storage unit, and all other costs that increase in proportion to installed collector area. The fixed costs include cost of pumps, fans, controls, extra ductwork, heat exchangers, the remainder of the storage unit, and all other costs that are

TABLE 6.4 WORKSHEET FOR EXAMPLE 6.1

F-CHART WORKSHEET 4

ECONOMIC PARAMETERS

H.	Annual mortgage interest rate	0.09 %/100
I.	Term of mortgage	20 Yrs.
J.	Down payment (as fraction of investment)	0.1 %/100
K.	Collector area dependent costs	200 $/m^2
L.	Area independent costs	1000 $
M.	Present cost of solar backup system fuel	8.33 $/GJ
N.	Present cost of conventional system fuel	8.33 $/GJ
O.	Efficiency of solar backup furnace	1.0 %/100
P.	Efficiency of conventional system furnace	1.0 %/100
Q.	Property tax rate (as fraction of investment)	0.0133 %/100
R.	Effective income tax bracket (state+federal-state x federal)	0.46 %/100
S.	Extra ins. & maint. costs (as fraction of investment)	0.01 %/100
T.	General inflation rate per year	0.06 %/100
**U.	Fuel inflation rate per year	0.10 %/100
V.	Discount rate (after tax return on best alternative investment)	0.08 %/100
W.	Term of economic analysis	20 Yrs.
X.	First year non-solar fuel expense (total, C5)(N.)/(P.)÷10^9	1693 $
*Y.	Depreciation lifetime	— Yrs.
Z.	Salvage value (as fraction of investment)	0 %/100
AA.	Table 6.2 with Yr = (W.), Column = (U.) and Row = (V.)	22.169
BB.	" (W.) " (T.) " (V.)	15.596
CC.	" MIN(I.,W.) " (H.) " (V.)	20.242
†*DD.	" MIN(W.,Y.) " (Zero) " (V.)	—
EE.	" (I.) " (Zero) " (H.)	9.129
FF.	" MIN(I.,W.) " (Zero) " (V.)	9.818
GG.	(FF.)/(EE.), Loan payment	1.076
HH.	(GG.)+(CC.)[(H.)-1/(EE.)], Loan interest	0.680
II.	(J.)+(1-J.)[(GG.)-(HH.)(R.)], Capital cost	0.787
JJ.	(S.)(BB.), I&M cost	0.156
KK.	(Q.)(BB)(1-R.), Property tax	0.112
LL.	(Z.)/(1+V.)$^{(W.)}$, Salvage value	—
*MM.	(R.)(DD.)(1-Z.)/(Y.), Depreciation	—
NN.	Other costs (see Section 6.9)	—
OO.	(II.)+(JJ.)+(KK.)-(LL.)+(NN.), Residential costs	1.055
PP.	(II.)+(JJ.)(1-R.)+(KK.)-(LL.)-(MM.)+(NN.)(1-R.), Commercial costs	—

**For other fuel inflation factors see Section 6.9.

*Commercial only.

†Straight line only. Use Tables 6.3A or 6.3B for other depreciation methods.

independent of the collector area. In addition, the fixed costs will include the difference in price, if any, between the conventional furnace and the solar system backup furnace. Although the backup furnace must be capable of carrying 100% of the design heating load, as does the conventional furnace, the two may be different and may even use different fuels. For this example we have assumed the area dependent costs to be $200 per square meter and the area independent costs to be $1000.

Lines M and N are the anticipated cost of fuel for the first year of operation. For this example we have assumed that electricity is to be used at $0.03/kW-hr which is equivalent to $8.33 per GJ ($8.79 per million BTU). This example assumes that electricity rates do not vary with the usage. Most utilities have a rate structure such that large consumers pay less per unit. Lately, some utilities have abandoned this type of rate structure and, in fact, some have even initiated inverted structures in which the consumer pays more per unit as he uses more units. All of these various schemes can be accomodated in calculating the first year's fuel bill.

Lines O and P are the efficiencies of the furnaces which are needed to calculate the amount of fuel that must be used to deliver a unit of heat to the load. For electricity, this efficiency is 100% but for an oil or gas furnace this efficiency is often as low as 50 or 60% (see Chapter 4).

In some states, the installation of solar collectors will increase the assessed valuation and result in an increased property tax. The property tax rate in some locations is based upon the fair market value while in other locations it is based upon some fixed fraction of the market value. Line Q is used to calculate the dollars paid in taxes on each dollar invested. In this example the tax rate is assumed to be 2% of the assessed value while the ratio of assessed value to actual cost is assumed to be 0.67; thus the tax rate as a fraction of investment is 0.0133. In some locations property tax credits are given for a solar investment. Such tax credits will be discussed later.

The majority of tax savings are in the form of income tax reductions. If the same deductions are allowed for state taxes as for federal taxes, then the effective income tax rate is the federal plus state tax brackets minus the product of the two. The reason that the sum is reduced by the product is that state taxes are deductable from federal taxes. For this problem we have assumed a federal bracket of 40% and a state bracket of 10% for an effective tax bracket on line R of 46%.

Line S is for insurance and maintenance costs for the solar system as a percentage of the initial cost. Insurance for a conventional home is on the order of 0.5% and we will assume this same rate holds for the solar heating system. Extra maintenance costs for the solar system are assumed to be 0.5% for a total entry on line S of 1%.

The general inflation rate has been on the order of 6% per year (with some years much higher) while fuel in recent years has increased at a rate of over 10%. For many years before the "energy crisis," fuel inflation was actually less that the general inflation rate; the future trends must be estimated by the user as he sees the situation. The two inflation rates are entered on lines T and U.

The market discount rate for line V represents the best alternative investment for the prospective owner of the solar system. For homeowners, the best investment might be long-term certificates of deposit which yield 0 to 2% more return than the general inflation rate. For a business, the rate may be 0 to 4% above the inflation rate for a low-growth company to 10 or even 20% for a high-growth company. The company accountant usually has a number that he uses in evaluating alternatives. For this problem, for a residence, we will use a discount rate of 8%, which is 2% above the assumed general inflation rate.

The term of the economic analysis for line W can be the expected life of the system, the term of the mortgage, the depreciation lifetime of the building or any other time period. For this example, we will use 20 years, the same as the mortgage period.

The fuel expense without solar for the first year is the annual heating load multiplied by the cost of fuel divided by the furnace efficiency. This is shown on line X.

The depreciation lifetime (line Y) is for commercial property only and is usually taken as the smallest time period allowed by the Internal Revenue Service. The salvage value in line Z is needed for both commercial and residential property. For commercial property, the amount that can be depreciated is 1-Z; it is to the owners advantage to take Z as small as possible. At the end of the life of the system, some salvage value may exist although the system has been depreciated to zero. It would then be necessary to include this "windfall" as income. It is also possible that in an actual system the salvage value is negative; it may actually cost the owner money to have the system removed. For a residence, the owner may conclude that the system has some resale value. Consequently, this value should be added to the life cycle savings. For this example we have taken a conserative approach and assumed the salvage value is zero.

Entries in lines AA through FF are from Table 6.2 with different values of the three parameters. The meaning of MIN(I., W.) on line CC is that the smaller time period shown in I or W must be used. Lines AA through CC represent the sum of the discounted (present) value of all yearly payments of an expense that is inflating. Line DD is used for depreciating commercial property. Lines EE through HH are needed in loan calculations. Line GG is the present value of the payments on one dollar borrowed for I years at an interest rate of H with a market discount rate of V. Line HH is the present value of all interest paid on the one dollar loan.

The terms II through NN represent life cycle values of various cash flows as a fraction of the initial investment. For example, line II for this problem is 0.787, which means that financing the purchase of the equipment for 20 years with the given values of income taxes, downpayment, and discount rate results in a cost that is only 78.7% of the purchaes price. Likewise, the life cycle insurance and maintenance cost is 15.6% of the solar investment. Line NN will be discussed in Section 6.9.

Lines OO and PP represent the total life cycle cost as a fraction of the investment for residential and commercial property. For this example, the total life cycle cost is 105.5% of the investment.

The economic calculations are worked out for four different collector areas on worksheet 5 as shown in Table 6.5. The first column is used for a solar system with zero collector area. The zero area solar system is used because a small collector area is needed for later interpolation to find the optimum system and the calculation of the thermal performance at zero area is trivial. The other three areas are chosen to bracket the optimum. For this problem we have chosen areas of 25, 50, and 75 square meters with the corresponding solar fractions from Figure 5.2.

The solar investment for row R3 is calculated from the collector unit area cost and from the fixed cost.

The first year's fuel expense for the zero area solar system is shown in the first column of row R4. This fuel bill is the total load times the cost of the solar backup fuel divided by the furnace efficiency. This fuel bill is reduced by

TABLE 6.5 WORKSHEET FOR EXAMPLE 6.1

F-CHART WORKSHEET 5

ECONOMIC ANALYSIS

R1. Collector Area (Worksheet 3)	0	25	50	75
R2. Fraction by Solar (Worksheet 3)	0	0.27	0.43	0.56
R3. Investment in Solar (K.)(R1.)+(L.)	1000	6000	11000	16000
R4. 1st Year Fuel Expense (Total, C5)(1-R2.)(M.)/(O.)÷10^9	1693	1236	965	745
R5. Fuel Savings (X.-R4.)(AA.)	0	10131	16139	21016
R6. Expenses (Residential) (OO.)(R3.)	1055	6330	11605	16880
R7. Expenses (Commercial) (PP.)(R3.)	—	—	—	—
R8. Savings (Residential) (R5.)-(R6.)	-1055	3801	4534	4136
R9. Savings (Commercial) (R5.)(1-R.)-(R7.)	—	—	—	—

the fraction supplied by solar for the three different systems. The life cycle fuel savings, given in row R5, are then the difference between the fuel expense without solar and with solar times the appropriate inflation-discount factor.

The life cycle expenses are shown in row R6 for residential property and in row R7 for commercial property. The final result is the solar savings shown in rows R8 and R9. Solar savings can be negative which means that it is actually a loss. The characteristic of all such savings calculations is that at zero area the savings are negative, the savings usually increase as the area increases and ultimately for large areas the savings again become negative, as shown in Figure 6.2. The economic optimum solar system is found by plotting collector area versus solar savings and finding the area that maximizes savings. For this example, the most economic collector area is about 50 square meters but positive savings are realized from 5 to over 100 square meters.

EXAMPLE 6.2 Solar Economics for a Business

A comparison between the solar savings for commercial and residential property is interesting. Table 6.6 gives the same parameters as Example 6.1, except that the parameters needed for a commercial building have been added. For commercial property, the solar equipment can be depreciated. In order to make this calculation we need to know the depreciation lifetime and the salvage value. The depreciation lifetime for this example is 20 years as shown on line Y and the salvage value, line Z, is 10% of the original investment. Line DD is the present value of all yearly depreciation deductions (straight line only). (If other depreciation schemes are to be used, such as

TABLE 6.6 WORKSHEET FOR EXAMPLE 6.2

F-CHART WORKSHEET 4
ECONOMIC PARAMETERS

H.	Annual mortgage interest rate	0.09 %/100
I.	Term of mortgage	20 Yrs.
J.	Down payment (as fraction of investment)	0.1 %/100
K.	Collector area dependent costs	200 $/m^2
L.	Area independent costs	1000 $
M.	Present cost of solar backup system fuel	8.33 $/GJ
N.	Present cost of conventional system fuel	8.33 $/GJ
O.	Efficiency of solar backup furnace	1.0 %/100
P.	Efficiency of conventional system furnace	1.0 %/100
Q.	Property tax rate (as fraction of investment)	0.0133 %/100
R.	Effective income tax bracket (state+federal-state x federal)	0.46 %/100
S.	Extra ins. & maint. costs (as fraction of investment)	0.01 %/100
T.	General inflation rate per year	0.06 %/100
**U.	Fuel inflation rate per year	0.10 %/100
V.	Discount rate (after tax return on best alternative investment)	0.08 %/100
W.	Term of economic analysis	20 Yrs.
X.	First year non-solar fuel expense (total, C5)(N.)/(P.)÷10^9	1693 $
*Y.	Depreciation lifetime	20 Yrs.
Z.	Salvage value (as fraction of investment)	0.10 %/100
AA.	Table 6.2 with Yr = (W.), Column = (U.) and Row = (V.)	22.169
BB.	" (W.) " (T.) " (V.)	15.596
CC.	" MIN(I.,W.) " (H.) " (V.)	20.242
†*DD.	" MIN(W.,Y.) " (Zero) " (V.)	9.818
EE.	" (I.) " (Zero) " (H.)	9.129
FF.	" MIN(I.,W.) " (Zero) " (V.)	9.818
GG.	(FF.)/(EE.), Loan payment	1.076
HH.	(GG.)+(CC.)[(H.)-1/(EE.)], Loan interest	0.680
II.	(J.)+(1-J.)[(GG.)-(HH.)(R.)], Capital cost	0.787
JJ.	(S.)(BB.), I&M cost	0.156
KK.	(Q.)(BB)(1-R.), Property tax	0.112
LL.	(Z.)/(1+V.)$^{(W.)}$, Salvage value	0.022
*MM.	(R.)(DD.)(1-Z.)/(Y.), Depreciation	0.203
NN.	Other costs (see Section 6.9)	—
OO.	(II.)+(JJ.)+(KK.)-(LL.)+(NN.), Residential costs	—
PP.	(II.)+(JJ.)(1-R.)+(KK.)-(LL.)-(MM.)+(NN.)(1-R.), Commercial costs	0.758

**For other fuel inflation factors see Section 6.9.
*Commercial only.
†Straight line only. Use Tables 6.3A or 6.3B for other depreciation methods.

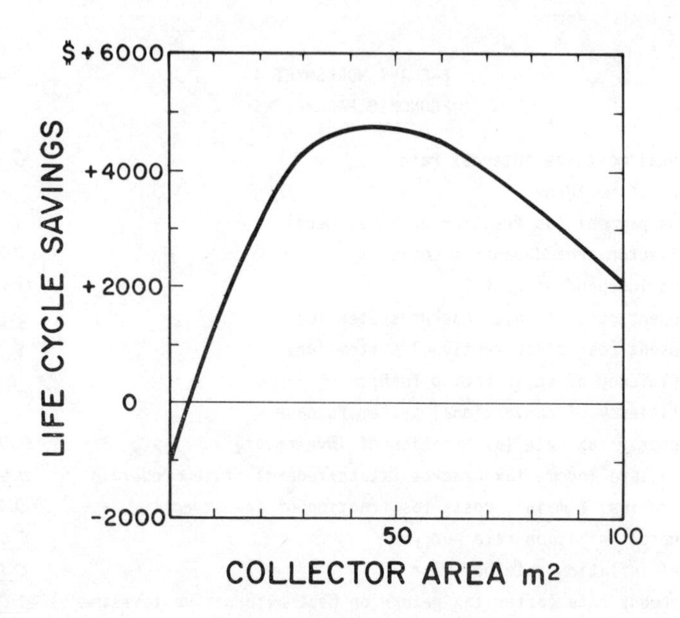

FIGURE 6.2
LIFE CYCLE SAVINGS AS A FUNCTION OF COLLECTOR AREA FOR RESIDENTIAL BUILDING OF EXAMPLE 6.1

sum-of-digits or double-declining balance, the factor DD is found in either Table 6.3A or 6.3B.)

Tablc 6.7 gives the 20-year savings with the commercial tax deductions. In row R9, the fuel sav.ings, from R5, must be reduced by 1 minus the income tax bracket since fuel savings resulting from installing a solar system is taxable income to the businessman (who has already deducted the cost of fuel as a business expense).

6.4 YEARLY SAVINGS

The previous examples gives the life cycle cost comparison but does not give the savings for individual years. Usually, the savings in the first few years are negative but if fuel prices increase faster than other expenses, the yearly savings can become positive. Because of concern over cash flow problems in early years it is sometimes interesting to

TABLE 6.7 WORKSHEET FOR EXAMPLE 6.2

F-CHART WORKSHEET 5

ECONOMIC ANALYSIS

R1. Collector Area (Worksheet 3)	0	25	50	75
R2. Fraction by Solar (Worksheet 3)	0	0.27	0.43	0.56
R3. Investment in Solar (K.)(R1.)+(L.)	1000	6000	11000	16000
R4. 1st Year Fuel Expense (Total, C5)(1-R2.)(M.)/(O.)÷10^9	1693	1236	965	745
R5. Fuel Savings (X.-R4.)(AA.)	0	10131	16139	21016
R6. Expenses (Residential) (OO.)(R3.)	—	—	—	—
R7. Expenses (Commercial) (PP.)(R3.)	758	4548	8338	12128
R8. Savings (Residential) (R5.)-(R6.)	—	—	—	—
R9. Savings (Commercial) (R5.)(1-R.)-(R7.)	-758	923	377	-779

calculate yearly savings. Worksheet 6 can be used to find yearly savings.

EXAMPLE 6.3 Yearly Savings

The first four year's savings for Example 6.1 are shown in row R23 of Table 6.8. As expected, major cash flow problems can exist in the first year.

The sum of all the yearly savings will not be equal to the life cycle savings unless each year's savings is discounted to the present. If the savings for year N is divided N-1 times by (1+d), where d is the market discount rate, the result is the discounted savings. The sum of all these yearly discounted savings will be equal to the life cycle savings. This calculation is performed in row R24 of worksheet 6. (The down payment in the first year should not be discounted.)

6.5 SAVINGS AS A FUNCTION OF FUEL COST

In Figure 6.3, we show solar savings for Example 6.1 as a function of collector area for four different initial electricty rates; $0.01, $0.02, $0.03, and $0.04 per kW-hr. (Electricty at $0.01 per kW-hr is equivalent to natural gas at $0.22 per 100 cubic feet with a 55% furnace efficiency.) As the initial fuel price increases, the solar savings at the optimum area goes from negative to positive. As expected, as the fuel price increases, both the savings and optimum area increase. Somewhat unexpected is the broad range of collector area that will give near optimum savings. Savings are positive for a fuel cost of $0.04 from 5 to well over 100 square meters. Thus, errors in finding the exact optimum are not critical. The 20 year savings are all above $10,000 for 45 to 100 square meters with a maximum of $11,000 at 70 square meters. Another interesting number is the "break even" electricity cost. If electricity were about

TABLE 6.8 WORKSHEET FOR EXAMPLE 6.3

F-CHART WORKSHEET 6

YEARLY SAVINGS FOR COLLECTOR AREA = 25 m^2

R10. Year (n) (first year=1)	1	2	3	4
†R11. Current Mortgage [(R11.)-(R14.)+(R18.)]	5400	5294	5178	5052
R12. Fuel Savings (X.-R4.)(1+U.)$^{n-1}$	457	502	553	608
R13. Down Payment (1st year only) (R3.)(J.)	600	—	—	—
R14. Mortgage Payment (1-J.)(R3.)/(EE.)	592	592	592	592
R15. Extra Insurance & Maintenance (S.)(R3.)(1+T.)$^{n-1}$	60	64	67	71
R16. Extra Property Tax (R3.)(Q.)(1+T.)$^{n-1}$	80	85	90	95
R17. Sum (R13.+R14.+R15.+R16.)	1332	741	749	758
R18. Interest on Mortgage (R11.)(H.)	486	476	466	455
R19. Tax Savings (R.)(R16.+R18.)	260	258	256	253
*R20. Depreciation (st. line) (R3.)(1-Z.)/(Y.)	—	—	—	—
*R21. Business Tax Savings (R.)(R20.+R15.-R12.)	—	—	—	—
R22. Salvage Value (R2.)(Z.) (Last Year Only)	—	—	—	—
R23. Solar Savings (R12.-R17.+R19.+R21.+R22.)	-615	19	60	103
**R24. Discounted Savings (R23.)/(1+V.)^n	-614	16	48	76

†For the first year use [(R3.)(1-J.)]; for subsequent years use equation with previous years values.

*Income producing property only.

**The down payment should not be discounted.

$0.018 per kW-hr, the 20-year savings would be zero with a collector of about 20 square meters.

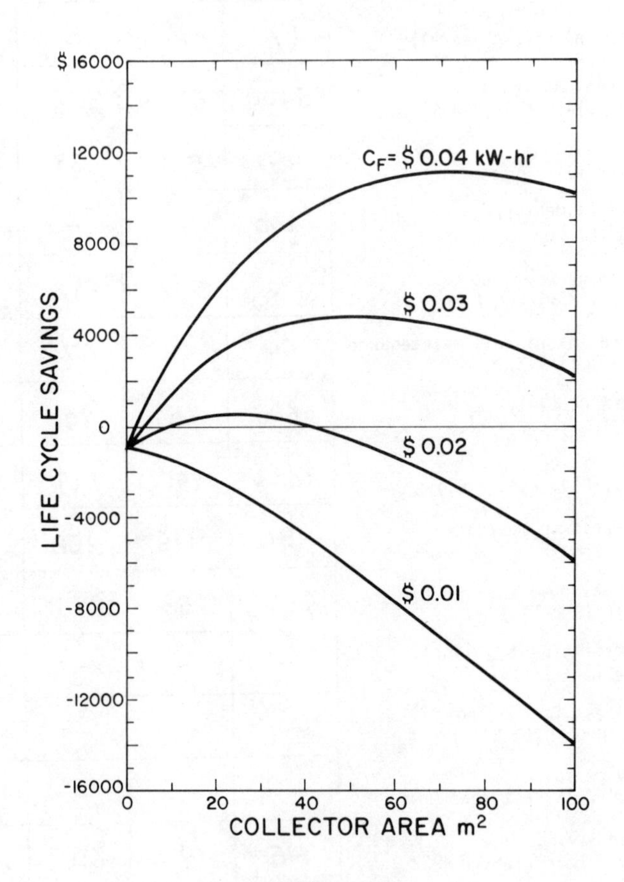

FIGURE 6.3
LIFE CYCLE SAVINGS AS A FUNCTION OF COLLECTOR AREA FOR FOUR DIFFERENT FUEL COSTS FOR BUILDING OF EXAMPLE 6.1 TREATED AS A RESIDENCE

Figure 6.4 shows savings as a function of collector area for four different initial fuel costs for the commercial property of Example 6.2. In this case the savings can be positive for any fuel cost above approximately $0.025 per kW-hr (instead of $0.018 as was the case for the residential building). The potential savings for commercial property are much smaller than for a residence. At an electricity cost of \$0.04 per kW-hr, the optimum 20-year savings are only \$3,400 compared to over \$11,000 maximum savings for the residential property. These differences are due entirely to tax laws. The current tax laws tend to make solar energy a more favorable choice for residential usage than for commercial usage.

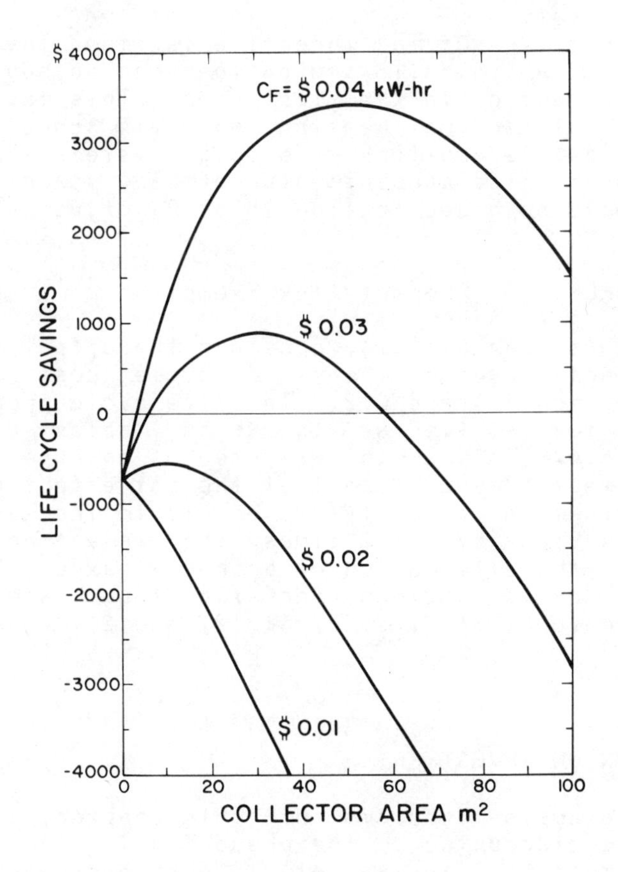

FIGURE 6.4
SAVINGS FOR FOUR DIFFERENT FUEL COSTS FOR BUILDING OF EXAMPLE 6.2 TREATED AS A BUSINESS

6.6 SPECIAL TAX CONSIDERATIONS

The federal government and many state governments are proposing tax credits to encourage solar energy utilization. The proposals range from exempting solar systems from the usual property tax, to giving a "bottom line tax credit." Since it is impossible to generalize such credits, it is necessary to treat each case separately.

Another type of tax incentive is the investment tax credit which allows a businessman an additional one-time deduction in the first year. This tax break is not allowed on heating and air conditioning equipment but is applicable to solar systems supplying process heat. The present value of any special tax credit would be added to line R8 or R9 of worksheet 5.

EXAMPLE 6.3 Property Tax Exemption

As an example, consider the effect of property taxes on the life cycle cost of Examples 6.1 and 6.2. The life cycle property tax is given as line KK of Tables 6.4 and 6.6. This amount was used as an expense in lines OO and PP so that the net effect of eliminating property taxes is to increase the savings by 0.112 times the investment. With the elimination of property taxes, the life cycle savings for the four areas increase by $112, $672, $1232, and $1792.

6.7 RETURN ON INVESTMENT

In the example problems of this chapter, all cash flows were discounted to the present using the "market discount rate." If the market discount had been zero, all life cycle savings would have been substantially increased. For the 25 square meter commercial system of Example 6.2, the discounted savings was $923. If a zero discount rate had been used, the savings would have been $4976. On the other hand, if a discount rate greater than 8% had been used, the life cycle savings would have been less than \$923. A 20% discount rate results in a loss of \$370. It is obvious that there is a discount rate that will give a zero life cycle savings. This discount rate is the "return on investment."

Finding the return on investment is a trial and error process. A particular design is chosen, usually the design producing the maximum savings. The savings calculations are repeated using different discount rates until zero life cycle savings is obtained. The process is best carried out graphically. Two discount rates are used initially, one giving positive savings and a second giving negative savings. The savings are plotted on a graph as a function of the discount rate. The two points are connected together by a straight line and the intersection with the zero savings axis is approximately the return on investment. If the two initial market discount rates are far from this new estimate of the return on investment, it may be necessary to compute a third point and draw a smooth curve through the three points. Figure 6.4 shows such a curve for the 25 square meter system of Example 6.2. The return on investment is seen to be 14%.

6.8 SECONDARY DESIGN VARIABLES

This chapter has used life cycle costs to examine only one design variable, the collector area. The guidelines given in Tables 1.1 and 1.2 for less important variables (i.e. secondary variables) can be expected to yield near optimum systems. However, if it is necessary to check to see if these recommendations are correct, it would be necessary to repeat both the thermal and economic calculations for a range of each secondary variable. Consider storage size as an example. For a number of different storage sizes we would have to find the collector area that yields the optimum savings. A curve is then drawn of optimum savings as a function of storage size. The optimum of this curve identifies the system that has both the optimum storage size and collector area. It is obvious that this is a time-consuming process that can only be justified in exceptional circumstances.

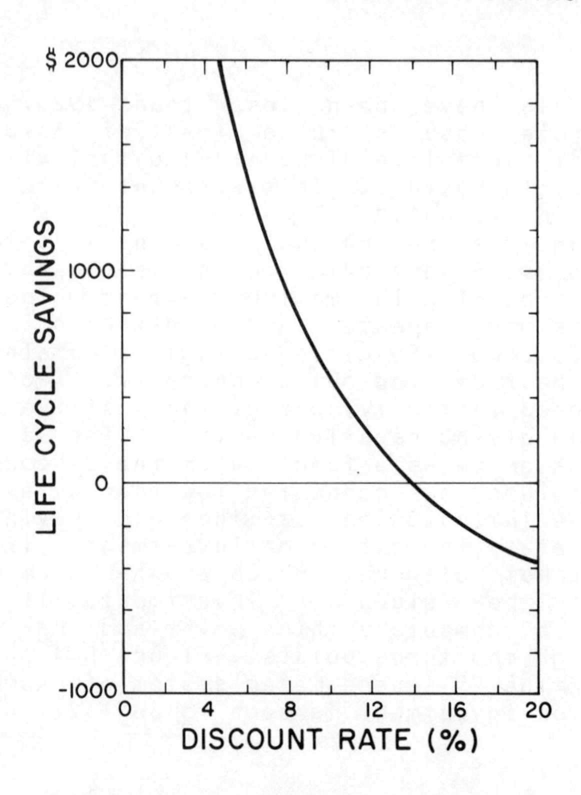

FIGURE 6.5
LIFE CYCLE SAVINGS AS A FUNCTION OF MARKET DISCOUNT RATE FOR 25 SQUARE METER SYSTEM OF EXAMPLE 6.2. INTERSECTION WITH 0 AXIS IS THE RETURN ON INVESTMENT.

6.9 IRREGULARLY VARYING COSTS

If costs vary with time in an irregular manner, that is, if they do not increase at a constant percentage per year, then the present value calculations of this chapter cannot be used directly. Although in principle it is possible to determine the present value of any arbitrary series of costs, detailed calculations are seldom justifiable becaues the exact nature of future cost escalations cannot be known with a high degree of certainty. Consequently, this section is concerned only with costs that vary in a reasonably well behaved manner.

If anticipated expenses are made at irregular intervals, the present worth of each of the expenses can be calculated by dividing an expense by $(1+d)^N$ where d is the discount rate and N is the year in which the payment is made. The sum of the present worths of all such expenses should then be subtracted from the savings in rows R8 or R9.

Section 6.2 shows how to use the inflation-discount function of Equation 6.7 to find the present worth of a maintenance or replacement cost that occurs each year. But what if a major expense occurs every few years? For example, a system with plastic glazing on the collector may have to have the covers replaced periodically. The present worth of such a series of payments can also be found using the inflation-discount function but with slightly different interpretations of the inflation and discount parameters. It is necessary to find the inflation rate and discount rate that is appropriate for the replacement time period based upon the assumed yearly values. With an inflation rate per year of i, an equivalent inflation rate of j per replacement time period, M, can be found from

$$j = (1+i)^M - 1 \qquad 6.8$$

In a similar manner, the discount rate of d per year is equivalent to a rate of c per M years, where c is given by

$$c = (1+d)^M - 1 \qquad 6.9$$

The life cycle replacement cost (LCRC) for line NN as a fraction of the initial investment is then equal to

$$LCRC = r \times F(M,j,c) \qquad 6.10$$

where r is the ratio of replacement cost at the time of the first replacement to the initial cost.

Another common type of an irregularly varying cost occurs when a fuel inflation rate of e per year is anticipated to level out after M years to a value of f per year until year N. The present worth of the fuel expense for the first M years, for each dollar of the first year's fuel payment, is F(M,e,d) from Equation 6.7. The cost of fuel at the end of year M, in terms of fuel prices at the end of the first year, is then $(1+e)^{M-1}$ and at the end of year M+1 is $(1+e)^{M-1}(1+f)$. The value, at the end of year M, of the N-M remaining fuel payments is then $(1+e)^{M-1}(1+f)F(N-M,f,d)$. The present worth of this amount is found by dividing by $(1+d)^M$. Therefore, the life cycle fuel cost (LCFC), starting with the first year's payment of one dollar is

$$LCFC = F(M,e,d) + \left[\frac{1+f}{1+e}\right]\left[\frac{1+e}{1+d}\right]^M \times F(N-M,f,d) \qquad 6.11$$

When appropriate, this equation can be used to find the entry for line AA of worksheet 4. The following example will illustrate these calculations.

EXAMPLE 6.4 Irregularly Varying Costs

If a plastic glazing on a collector is to be replaced every 5 years at a cost of materials and labor that is expected to be 20% of the initial investment (in terms of today's cost), what is the 20-year life cycle factor for line NN of Worksheet 4? Assume general inflation is 6% and the discount rate is 8%. Also, if fuel increases at 12% per year for the first 6 years and at 8% thereafter, what is the 20-year life cycle fuel factor for line AA of worksheet 4?

For this problem the covers must be repaced three times; at the end of years 5, 10 and 15. A replacement should not be considered in year 20. At the end of year 5 the replacement cost will be $0.20 \times \cdot (1.06)^5$ or 26.8% of the initial cost. An inflation rate of 6% per year is equivalent, from Equation 6.8, to 34% per 5 years and the 8% yearly discount rate is equivalent to 47% per 5 years from Equation 6.9. Therefore, the life cycle cost of the three cover replacements, as a fraction of the initial investment, is $0.268 \times F(3,0.34,0.47) = 0.268 \times 1.866 = 0.499$. This value would be entered on line NN of Worksheet 4.

The same results can be obtained by treating each cover replacement individually. The three cover replacement costs are anticipated to be (as a fraction of initial investment) 0.268, 0.358, and 0.479. The first replacement must be discounted 5 years, the second 10 years and the third 15 years. The present worth of the three costs are then 0.182, 0.166, and 0.151. The sum of these three costs is 0.499, the same as the previous calculation.

The life cycle fuel cost is calculated using Equation 6.11. Substituting in the appropriate values

$$\mathrm{LCFC} = F(6, 0.12, 0.08) + \left[\frac{1.09}{1.12}\right]\left[\frac{1.12}{1.08}\right]^6 \times F(14, 0.09, 0.08) = 22.768$$

This number should be entered on line AA of Worksheet 4.

6.10 SUMMARY

In this chapter we have used the results of the thermal analysis to find the collector area that maximizes the discounted life cycle savings. Solar savings were discounted to take into account the time value of money. Income tax deductions were shown to be of major importance. Present tax laws were shown to favor residential adoption of solar heating over commericial adoption.

Finally, we must caution the reader not to assume the parameters used in the illustrative examples are valid for his location. The values of the economic parameters used in this chapter were chosen to be representive of those encountered in many locations, but clearly are not generally applicable. The reader should determine local costs for his own economic calculations.

APPENDIX 1
HEAT EXCHANGER EFFECTIVENESS

In a liquid-based solar heating system, such as shown in Figure 1.1, there are often two or more heat exchangers. A liquid-to-liquid heat exchanger is needed to transfer heat from the solar collectors to the storage tank when a separate flow circuit containing antifreeze and/or corrosion inhibitors is used in the collectors. A second liquid-to-liquid heat exchanger may be used in some systems to transfer heat from the storage tank to the domestic water preheat tank. A water-to-air (usually crossflow) heat exchanger is needed to transfer heat from the storage tank to the building air.

The performance of the solar heating system is affected by the performance of the heat exchangers in the system. In this text, heat exchanger performance is represented in terms of the heat exchanger effectiveness, ε, and the minimum capacitance rate (i.e., mass flowrate x specific heat) in the heat exchanger. As a result, it is necessary to be able to determine the effectiveness of a heat exchanger from the performance data ordinarily supplied by the manufacturer.

The effectiveness of a heat exchanger, ε, is the ratio of the actual rate of heat transfer to the maximum possible heat transfer rate.

$$\varepsilon = \frac{\text{ACTUAL HEAT TRANSFER RATE}}{\text{MAXIMUM HEAT TRANSFER RATE}} \qquad \text{A1.1}$$

The effectiveness will always have a value between 0 and 1. In a specific heat exchanger, both the actual and the maximum heat transfer rates will vary as the temperatures of the entering flow streams change. The advantage of the effectiveness concept is that the ratio of the actual to maximum heat transfer rates is nearly constant, independent of temperature variations, if the mass flowrates in the heat exchanger are constant.

The maximum heat transfer rate is the product of the minimum capacitance rate, C_{min}, and the difference between the temperatures of the entering hot and cold flow streams, T_h and T_c respectively.

$$\text{MAXIMUM HEAT TRANSFER RATE} = C_{min} \times (T_h - T_c) \qquad \text{A1.2}$$

In the water-to-air heat exchanger in which space heat is transferred from the water storage tank, the minimum capacitance rate, C_{min}, is usually the capacitance rate of the air. In this case, C_{min} would be the product of the mass flowrate and the specific heat of the air. In the collector-tank heat exchanger, C_{min} is often the capacitance rate of the antifreeze solution flowing in the solar collectors because the mass flowrates of the water and antifreeze solutions are often equal, but the specific heat of the antifreeze solution is lower than that of water.

The actual heat transfer rate is a function of the heat exchanger size and design, and the capacitance rates and temperatures of the flow streams. The actual heat transfer rate, or equivalently, the temperature rise of either flow stream, for a specific heat exchanger operating at known conditions is ordinarily available from tables or charts supplied by the manufacturer.

EXAMPLE A1.1 Effectiveness of a Load Heat Exchanger

A crossflow water-to-air heat exchanger is to be used in a solar heating system to supply the space heating load from hot water in a storage tank. The water flowrate will be 0.694 liters per second (11 gpm) and the air flowrate will be 520 liters of air at standard conditions per second (1100 cfm). Data from the manufacturer indicate that the rate of heat transfer for this heat exchanger at these flow conditions will be 16660 W (56860 BTU/hr) when the entering air temperature is 22 C and the entering water temperature is 60 C. Calculate the effectiveness of this heat exchanger.

The capacitance rate of the water, C_w, is

$$C_w = (0.694\ \text{l/s}) \times (1\ \text{kg/l}) \times (4190\ \text{J/kg-C})$$
$$= 2910\ \text{W/C}$$

The capacitance rate of the air, C_a, is

$$C_a = (520\ l/s) \times (0.001204\ kg/l) \times (1010\ J/kg\text{-}C) = 632\ W/C$$

The minimum capacitance rate, C_{min}, is thus that of the air.

The maximum possible heat transfer rate is the product of C_{min} and the temperature difference between the entering flow streams, which is (60 C - 22 C) or 38 C.

MAXIMUM HEAT TRANSFER RATE = 632 W/C X 38 C = 24016 W

The effectiveness of the load heat exchanger, ε_L, is the ratio of the actual to the maximum heat transfer rates. Thus,

$$\varepsilon_L = (16660\ W)/(24016\ W) = 0.69$$

APPENDIX 2
METEOROLOGICAL DATA

Presented in the following tables are meteorological data for 171 locations in the United States and Canada. The tables are arranged in alphabetical order by city name, with the United States cities first, followed by the Canadian cities.

The data include: $\overline{H}$, the monthly average of daily radiation on a horizontal surface, in megajoules per square meter (MJ/m^2); $\overline{K}_T$, the ratio of average daily radiation to average daily extraterrestrial radiation; $\overline{T}_a$, the monthly mean ambient temperature (C); and DD, the average number of degree-days in the month (C-days).

The data in the tables have been compiled from several sources, as listed below. It should be noted that there may be significant errors in the radiation data arising from past instrumental difficulties; radiation measurements are now being upgraded in the United States network for solar radiation measurements. When more recent data are available to provide a basis for revised averages, they should be used.

Klein et al. (1977), in the last reference in the list below, discuss in some detail, the calculation of average radiation on tilted surfaces of various inclinations. Included in the report are calculated estimates of radiation on surfaces with slopes of 20° to 90°, and at azimuth angles of 0° to 45°, based on the data included in this appendix. There are uncertainties in the calculation methods, and these methods will undoubtedly be refined as more reliable data become available.

ASHRAE GUIDE AND DATA BOOK, SYSTEMS, American Society of Heating, Refrigerating and Air Conditioning Engineers, New York (1973)

Beck E.J. and Field R.L., Solar Heating of Buildings and Domestic Hot Water, Technical Report R835, Naval Facilities Engineering Command (1976)

Lof, G.O.G., Duffie, J.A., and Smith, C.O., World Distribution of Solar Radiation, Report No. 21, Engineering Experiment Station, University of Wisconsin - Madison (1966)

U.S. Dept. of Commerce, Climactic Atlas of the United States, Environmental Data Service, Reprinted by the National Oceanic and Atmospheric Administration (1974)

U.S. Dept. of Commerce, Monthly Normal of Temperature, Precipitation, and Heating and Cooling Degree-Days (1941-1970), National Oceanic and Atmospheric Administration, Climatography of the United States No. 81

Klein, S.A., Beckman, W.A., and Duffie, J.A., Monthly Average Solar Radiation on Inclined Surfaces for 171 North American Cities, Report No. 44, Engineering Experiment Station, University of Wisconsin - Madison (1977)

	JAN	FEB	MAR	APR	MAY	JUNE	JULY	AUG	SEP	OCT	NOV	DEC
ABILENE	TX (LAT. 32.2)											
$\overline{H}$ MJ/m^2	11.47	14.74	18.42	22.11	24.24	27.05	25.41	23.74	20.43	16.24	12.31	10.51
$\overline{K}_T$	.58	.60	.61	.61	.61	.66	.63	.64	.63	.62	.59	.57
$\overline{T}_a$ C	6.0	9.0	12.0	18.0	22.0	27.0	29.0	29.0	24.0	19.0	12.0	8.0
DD C-DAY	367.	266.	197.	58.	6.	0.	0.	0.	0.	49.	187.	321.
ALBANY	NY (LAT. 42.4)											
$\overline{H}$ MJ/m^2	5.19	7.65	14.34	15.22	18.44	25.22	22.37	18.65	13.05	10.75	9.20	5.86
$\overline{K}_T$	.38	.40	.55	.45	.47	.61	.56	.52	.45	.51	.62	.48
$\overline{T}_a$ C	-5.0	-4.0	.0	8.0	14.0	19.0	22.0	21.0	17.0	10.0	4.0	-2.0
DD C-DAY	749.	646.	544.	302.	141.	22.	5.	12.	75.	234.	423.	673.
ALBUQUERQUE	NM (LAT. 35.0)											
$\overline{H}$ MJ/m^2	12.88	16.31	21.41	26.35	28.77	30.90	28.86	26.60	23.67	18.69	14.13	11.75
$\overline{K}_T$	.71	.71	.73	.74	.73	.75	.72	.72	.75	.75	.74	.71
$\overline{T}_a$ C	1.0	4.0	7.0	12.0	17.0	22.0	25.0	23.0	20.0	13.0	6.0	1.0
DD C-DAY	513.	389.	331.	157.	32.	0.	0.	0.	4.	121.	342.	496.
AMARILLO	TX (LAT. 35.1)											
$\overline{H}$ MJ/m^2	11.93	15.24	19.17	23.61	25.83	27.38	26.75	25.12	21.23	16.83	12.98	10.72
$\overline{K}_T$	.66	.66	.65	.67	.65	.67	.66	.68	.67	.68	.68	.65
$\overline{T}_a$ C	2.0	4.0	8.0	14.0	19.0	24.0	26.0	25.0	21.0	15.0	8.0	4.0
DD C-DAY	499.	393.	334.	153.	45.	6.	0.	0.	11.	114.	312.	457.
AMES	IA (LAT. 42.0)											
$\overline{H}$ MJ/m^2	7.28	10.58	13.67	16.85	20.07	22.62	22.41	19.24	15.35	11.46	7.82	5.98
$\overline{K}_T$	.53	.55	.52	.50	.51	.55	.56	.54	.53	.54	.52	.48
$\overline{T}_a$ C	-7.0	-4.0	.0	9.0	15.0	20.0	23.0	22.0	17.0	11.0	2.0	-4.0
DD C-DAY	794.	639.	539.	260.	106.	18.	0.	8.	58.	206.	463.	699.
AMHERST	MA (LAT. 42.1)											
$\overline{H}$ MJ/m^2	4.85	7.40	12.55	14.51	18.02	21.49	21.58	18.40	13.80	10.45	6.40	5.19
$\overline{K}_T$	.35	.39	.48	.43	.46	.52	.54	.51	.48	.49	.43	.42
$\overline{T}_a$ C	-4.0	-3.0	1.0	8.0	14.0	19.0	21.0	20.0	16.0	11.0	4.0	-2.0
DD C-DAY	713.	611.	515.	300.	139.	23.	4.	12.	68.	221.	403.	644.

	JAN	FEB	MAR	APR	MAY	JUNE	JULY	AUG	SEP	OCT	NOV	DEC
ANNAPOLIS MD (LAT. 38.6)												
$\overline{H}$ MJ/m^2	7.33	10.17	14.24	17.54	20.43	23.32	22.69	19.64	16.03	12.31	7.91	6.49
$\overline{K}_T$	.46	.48	.51	.51	.52	.56	.56	.54	.53	.53	.46	.45
$\overline{T}_a$ C	1.0	2.0	6.0	12.0	17.0	22.0	24.0	23.0	20.0	14.0	8.0	2.0
DD C-DAY	526.	454.	376.	183.	58.	0.	0.	0.	16.	137.	293.	484.
ANNETTE AK (LAT. 55.0)												
$\overline{H}$ MJ/m^2	2.64	4.81	9.88	15.24	18.30	18.34	18.34	14.28	10.80	5.11	2.47	1.72
$\overline{K}_T$	.43	.42	.51	.52	.49	.45	.47	.44	.47	.37	.34	.36
$\overline{T}_a$ C	.0	2.0	3.0	5.0	9.0	11.0	15.0	14.0	12.0	8.0	4.0	2.0
DD C-DAY	527.	465.	468.	381.	281.	187.	143.	124.	196.	315.	410.	499.
APALACHICOLA FL (LAT. 29.4)												
$\overline{H}$ MJ/m^2	12.25	15.22	18.44	23.00	25.47	24.71	22.62	21.20	19.24	17.48	13.93	11.04
$\overline{K}_T$	.57	.59	.59	.63	.64	.61	.57	.56	.58	.63	.62	.55
$\overline{T}_a$ C	12.0	13.0	15.0	19.0	23.0	26.0	27.0	27.0	26.0	21.0	16.0	13.0
DD C-DAY	193.	144.	100.	18.	0.	0.	0.	0.	0.	9.	85.	177.
ASHEVILLE NC (LAT. 35.3)												
$\overline{H}$ MJ/m^2	9.21	12.48	16.03	20.51	23.28	23.78	23.24	21.56	18.25	14.86	10.47	8.46
$\overline{K}_T$	.51	.54	.55	.58	.59	.58	.58	.58	.58	.60	.55	.51
$\overline{T}_a$ C	3.0	4.0	8.0	13.0	18.0	21.0	23.0	23.0	19.0	14.0	8.0	4.0
DD C-DAY	467.	398.	329.	155.	56.	8.	0.	0.	28.	149.	312.	453.
ASTORIA OR (LAT. 46.1)												
$\overline{H}$ MJ/m^2	3.85	6.52	11.17	15.51	20.49	20.16	22.37	19.07	15.01	8.82	4.77	3.26
$\overline{K}_T$	.34	.39	.46	.48	.53	.49	.56	.55	.55	.47	.38	.33
$\overline{T}_a$ C	5.0	6.0	6.0	8.0	11.0	13.0	15.0	15.0	14.0	11.0	8.0	5.0
DD C-DAY	420.	333.	355.	287.	219.	142.	91.	84.	112.	210.	308.	382.
ATLANTA GA (LAT. 33.4)												
$\overline{H}$ MJ/m^2	9.53	11.88	15.77	20.24	22.37	23.17	22.50	20.99	17.23	14.64	11.08	8.41
$\overline{K}_T$	.50	.50	.53	.56	.56	.56	.56	.56	.54	.57	.55	.48
$\overline{T}_a$ C	6.0	7.0	11.0	16.0	20.0	24.0	25.0	25.0	22.0	17.0	11.0	7.0
DD C-DAY	389.	311.	246.	80.	15.	0.	0.	0.	4.	76.	227.	371.

	JAN	FEB	MAR	APR	MAY	JUNE	JULY	AUG	SEP	OCT	NOV	DEC
ATLANTIC CITY NJ (LAT. 39.3)												
$\overline{H}$ MJ/m^2	7.41	10.68	16.12	18.09	20.64	23.99	23.78	19.97	16.41	12.64	8.83	6.66
$\overline{K}_T$	.48	.51	.59	.52	.52	.58	.59	.55	.55	.56	.53	.47
$\overline{T}_a$ C	2.0	2.0	5.0	10.0	15.0	21.0	23.0	23.0	20.0	15.0	9.0	3.0
DD C-DAY	520.	461.	407.	242.	100.	8.	0.	0.	13.	111.	277.	470.
BALTIMORE MD (LAT. 39.1)												
$\overline{H}$ MJ/m^2	7.33	10.17	14.24	17.54	20.43	23.32	22.69	19.64	16.03	12.31	7.91	6.49
$\overline{K}_T$	.47	.49	.52	.51	.52	.56	.56	.54	.53	.54	.47	.46
$\overline{T}_a$ C	1.0	2.0	6.0	12.0	18.0	22.0	25.0	24.0	20.0	14.0	8.0	2.0
DD C-DAY	544.	470.	382.	189.	61.	0.	0.	0.	15.	139.	315.	512.
BARROW AK (LAT. 71.2)												
$\overline{H}$ MJ/m^2	.13	1.67	7.99	16.85	20.95	22.87	17.73	10.62	4.81	1.72	.29	.00
$\overline{K}_T$	.00	.79	.81	.76	.60	.54	.46	.39	.34	.42	.00	.00
$\overline{T}_a$ C	-25.0	-28.0	-26.0	-18.0	-7.0	.0	3.0	3.0	.0	-9.0	-18.0	-24.0
DD C-DAY	1398.	1296.	1371.	1080.	803.	547.	446.	467.	575.	833.	1095.	1271.
BETHEL AK (LAT. 60.5)												
$\overline{H}$ MJ/m^2	1.55	4.68	11.79	18.57	19.19	18.78	15.47	10.62	8.32	4.89	1.88	.96
$\overline{K}_T$	.49	.57	.72	.69	.53	.46	.40	.35	.41	.47	.44	.47
$\overline{T}_a$ C	-14.0	-13.0	-11.0	-4.0	4.0	10.0	12.0	11.0	7.0	-1.0	-8.0	-15.0
DD C-DAY	1057.	883.	919.	652.	448.	223.	177.	219.	340.	579.	797.	1037.
BIG SPRING TX (LAT. 32.1)												
$\overline{H}$ MJ/m^2	11.21	14.39	19.49	24.38	23.92	24.80	23.08	19.57	21.87	16.14	12.17	10.87
$\overline{K}_T$	.57	.58	.64	.68	.60	.61	.57	.52	.67	.61	.58	.59
$\overline{T}_a$ C	6.0	9.0	12.0	18.0	22.0	26.0	28.0	28.0	24.0	18.0	12.0	7.0
DD C-DAY	362.	260.	179.	50.	0.	0.	0.	0.	0.	48.	212.	329.
BILLINGS MT (LAT. 45.5)												
$\overline{H}$ MJ/m^2	6.62	9.92	15.03	19.09	22.65	25.62	26.63	23.32	17.75	10.68	7.24	5.57
$\overline{K}_T$	.56	.58	.61	.58	.58	.62	.67	.67	.65	.55	.56	.54
$\overline{T}_a$ C	-6.0	-3.0	.0	7.0	12.0	17.0	22.0	21.0	15.0	10.0	2.0	-3.0
DD C-DAY	742.	585.	558.	340.	185.	73.	6.	8.	123.	271.	488.	658.

	JAN	FEB	MAR	APR	MAY	JUNE	JULY	AUG	SEP	OCT	NOV	DEC
BINGHAMPTON NY (LAT. 42.1)												
$\overline{H}$ MJ/m^2	5.82	8.54	12.48	15.99	20.39	23.32	22.90	19.72	15.24	10.80	6.03	4.77
$\overline{K}_T$	.42	.45	.48	.47	.52	.56	.57	.55	.53	.51	.40	.39
$\overline{T}_a$ C	-6.0	-5.0	.0	7.0	13.0	18.0	21.0	20.0	16.0	10.0	3.0	-4.0
DD C-DAY	741.	657.	581.	338.	178.	42.	12.	22.	96.	253.	447.	682.
BIRMINGHAM AL (LAT. 33.3)												
$\overline{H}$ MJ/m^2	8.58	11.85	15.32	20.56	23.40	23.57	22.94	21.18	18.09	15.03	10.38	8.12
$\overline{K}_T$	.45	.49	.51	.57	.59	.57	.57	.57	.56	.58	.51	.46
$\overline{T}_a$ C	7.0	8.0	12.0	17.0	21.0	25.0	27.0	26.0	23.0	17.0	11.0	7.0
DD C-DAY	363.	287.	216.	64.	11.	0.	0.	0.	3.	76.	217.	341.
BISMARCK ND (LAT. 46.5)												
$\overline{H}$ MJ/m^2	6.61	10.50	14.68	18.78	23.04	24.55	25.59	21.66	15.98	11.42	6.73	5.19
$\overline{K}_T$	.59	.63	.61	.58	.60	.59	.64	.62	.59	.61	.54	.54
$\overline{T}_a$ C	-13.0	-11.0	-3.0	6.0	12.0	17.0	21.0	20.0	14.0	7.0	-1.0	-9.0
DD C-DAY	978.	801.	687.	367.	188.	68.	10.	19.	140.	313.	602.	851.
BLUE HILL MA (LAT. 42.1)												
$\overline{H}$ MJ/m^2	6.52	8.99	12.71	15.85	19.70	21.62	20.91	18.15	14.72	10.41	6.61	5.39
$\overline{K}_T$	.47	.47	.49	.47	.50	.52	.52	.51	.51	.49	.44	.44
$\overline{T}_a$ C	-3.0	-3.0	1.0	7.0	13.0	18.0	21.0	20.0	16.0	11.0	5.0	-1.0
DD C-DAY	654.	585.	520.	322.	148.	38.	0.	12.	59.	212.	383.	603.
BOISE ID (LAT. 43.3)												
$\overline{H}$ MJ/m^2	5.94	9.74	14.18	20.32	24.55	26.72	27.98	23.80	19.07	13.13	7.57	5.14
$\overline{K}_T$	.46	.53	.55	.61	.63	.65	.70	.67	.67	.64	.53	.44
$\overline{T}_a$ C	-1.0	1.0	5.0	9.0	14.0	18.0	23.0	22.0	17.0	11.0	4.0	.0
DD C-DAY	618.	474.	401.	243.	136.	45.	0.	0.	73.	231.	440.	565.
BOSTON MA (LAT. 42.2)												
$\overline{H}$ MJ/m^2	5.81	8.28	12.25	15.22	19.74	20.87	20.74	17.77	14.26	9.95	6.06	4.98
$\overline{K}_T$	.42	.43	.47	.45	.50	.50	.52	.50	.49	.47	.41	.41
$\overline{T}_a$ C	-1.0	.0	3.0	9.0	15.0	20.0	23.0	22.0	18.0	13.0	7.0	1.0
DD C-DAY	604.	540.	470.	285.	116.	20.	0.	5.	33.	176.	335.	546.

		JAN	FEB	MAR	APR	MAY	JUNE	JULY	AUG	SEP	OCT	NOV	DEC
BOULDER	CO (LAT. 40.0)												
$\bar{H}$	MJ/m^2	8.41	11.22	16.79	19.26	19.26	21.98	21.77	18.38	17.25	12.98	9.29	7.62
$\bar{K}_T$		.56	.55	.62	.56	.49	.53	.54	.51	.58	.58	.57	.56
$\bar{T}_a$	C	.0	1.0	3.0	9.0	14.0	19.0	23.0	22.0	17.0	12.0	5.0	2.0
DD	C-DAY	551.	459.	449.	268.	131.	49.	3.	0.	77.	204.	383.	503.
BROWNSVILLE	TX (LAT. 25.5)												
$\bar{H}$	MJ/m^2	12.00	14.05	16.81	19.15	23.25	25.26	25.89	23.21	19.45	16.98	11.88	10.58
$\bar{K}_T$		.51	.50	.51	.52	.59	.63	.65	.61	.57	.58	.48	.47
$\bar{T}_a$	C	15.0	17.0	20.0	23.0	26.0	28.0	28.0	28.0	27.0	24.0	19.0	16.0
DD	C-DAY	125.	84.	49.	0.	0.	0.	0.	0.	0.	3.	19.	81.
CAPE HATTERAS	NC (LAT. 35.2)												
$\bar{H}$	MJ/m^2	10.20	13.26	18.07	23.88	26.56	26.97	26.30	23.29	19.74	15.10	11.88	9.03
$\bar{K}_T$		.57	.58	.62	.67	.67	.66	.65	.63	.63	.61	.62	.54
$\bar{T}_a$	C	8.0	8.0	10.0	14.0	19.0	23.0	25.0	25.0	23.0	18.0	13.0	9.0
DD	C-DAY	339.	299.	254.	104.	26.	0.	0.	0.	0.	42.	154.	298.
CARIBOU	ME (LAT. 46.5)												
$\bar{H}$	MJ/m^2	5.73	9.62	15.35	16.73	19.82	20.07	21.29	18.82	13.93	8.78	4.60	4.43
$\bar{K}_T$		.52	.58	.64	.52	.51	.49	.53	.54	.52	.47	.37	.46
$\bar{T}_a$	C	-11.0	-10.0	-4.0	3.0	10.0	15.0	18.0	17.0	12.0	7.0	.0	-8.0
DD	C-DAY	939.	817.	727.	477.	260.	102.	43.	64.	187.	379.	580.	853.
CHARLESTON	SC (LAT. 32.5)												
$\bar{H}$	MJ/m^2	10.58	12.67	16.39	21.54	23.00	23.42	21.87	20.74	17.06	14.34	11.92	9.03
$\bar{K}_T$		.54	.52	.54	.60	.58	.57	.54	.56	.53	.55	.58	.50
$\bar{T}_a$	C	10.0	10.0	14.0	18.0	22.0	25.0	27.0	26.0	24.0	19.0	14.0	10.0
DD	C-DAY	271.	216.	162.	30.	0.	0.	0.	0.	0.	33.	157.	262.
CHARLOTTE	NC (LAT. 35.1)												
$\bar{H}$	MJ/m^2	9.29	12.39	16.20	21.39	23.07	24.45	23.65	21.69	18.21	14.86	10.51	8.54
$\bar{K}_T$		.52	.54	.55	.60	.58	.59	.59	.59	.58	.60	.55	.51
$\bar{T}_a$	C	6.0	7.0	10.0	16.0	20.0	24.0	26.0	25.0	22.0	16.0	11.0	6.0
DD	C-DAY	394.	327.	256.	81.	19.	0.	0.	0.	6.	84.	233.	388.

	JAN	FEB	MAR	APR	MAY	JUNE	JULY	AUG	SEP	OCT	NOV	DEC
CHATTANOOGA TN (LAT. 35.0)												
$\overline{H}$ MJ/m^2	8.12	11.35	14.49	19.51	22.57	23.32	22.90	21.02	17.84	13.94	9.46	7.58
$\overline{K}_T$	.45	.49	.49	.55	.57	.57	.57	.57	.57	.56	.49	.45
$\overline{T}_a$ C	5.0	6.0	10.0	16.0	20.0	24.0	26.0	26.0	22.0	16.0	9.0	5.0
DD C-DAY	427.	347.	268.	92.	28.	0.	0.	0.	5.	101.	268.	410.
CHICAGO IL (LAT. 41.6)												
$\overline{H}$ MJ/m^2	7.15	9.70	13.63	16.31	20.78	23.13	22.04	20.32	16.06	11.08	6.57	5.48
$\overline{K}_T$	.51	.50	.52	.48	.53	.56	.55	.57	.55	.52	.43	.43
$\overline{T}_a$ C	-3.0	-2.0	3.0	10.0	16.0	21.0	24.0	23.0	19.0	13.0	5.0	-1.0
DD C-DAY	701.	585.	486.	252.	116.	14.	0.	4.	32.	176.	410.	629.
CLEVELAND OH (LAT. 41.2)												
$\overline{H}$ MJ/m^2	5.19	7.53	13.05	15.77	21.87	23.38	23.04	20.62	15.72	11.00	5.90	4.81
$\overline{K}_T$	.36	.38	.49	.46	.56	.57	.57	.57	.54	.51	.38	.37
$\overline{T}_a$ C	-2.0	-1.0	2.0	9.0	15.0	20.0	22.0	21.0	18.0	12.0	5.0	.0
DD C-DAY	656.	577.	498.	278.	136.	22.	5.	9.	53.	197.	390.	598.
COLUMBIA MO (LAT. 38.6)												
$\overline{H}$ MJ/m^2	7.53	10.45	14.39	18.11	22.21	23.88	24.00	22.00	18.73	13.55	9.28	7.07
$\overline{K}_T$	.47	.49	.52	.52	.56	.58	.60	.60	.62	.59	.54	.49
$\overline{T}_a$ C	-1.0	.0	6.0	12.0	18.0	23.0	25.0	24.0	20.0	14.0	6.0	.0
DD C-DAY	598.	486.	398.	180.	67.	7.	0.	0.	30.	139.	362.	537.
COLUMBUS OH (LAT. 40.0)												
$\overline{H}$ MJ/m^2	5.39	8.28	12.38	16.43	20.41	23.50	22.67	19.95	17.65	11.96	7.44	5.52
$\overline{K}_T$	.36	.41	.46	.48	.52	.57	.56	.55	.59	.54	.46	.41
$\overline{T}_a$ C	-1.0	.0	4.0	11.0	16.0	21.0	23.0	22.0	18.0	12.0	5.0	.0
DD C-DAY	604.	527.	449.	237.	95.	15.	0.	3.	47.	193.	397.	577.
CORPUS CHRISTI TX (LAT. 27.5)												
$\overline{H}$ MJ/m^2	10.97	13.82	17.29	19.85	23.49	25.29	26.33	23.36	19.68	17.08	11.93	10.05
$\overline{K}_T$	.49	.51	.54	.54	.59	.63	.66	.62	.58	.60	.51	.47
$\overline{T}_a$ C	13.0	15.0	18.0	22.0	25.0	27.0	29.0	29.0	27.0	23.0	18.0	15.0
DD C-DAY	169.	111.	67.	0.	0.	0.	0.	0.	0.	4.	45.	122.

	JAN	FEB	MAR	APR	MAY	JUNE	JULY	AUG	SEP	OCT	NOV	DEC
CORVALLIS OR (LAT. 44.3)												
$\bar{H}$ MJ/m^2	4.22	5.81	11.75	16.89	21.24	24.30	28.02	22.87	16.69	9.83	5.86	3.39
$\bar{K}_T$	.34	.33	.47	.51	.55	.59	.70	.65	.60	.49	.43	.31
$\bar{T}_a$ C	3.0	6.0	7.0	10.0	13.0	16.0	18.0	18.0	16.0	11.0	7.0	5.0
DD C-DAY	451.	341.	336.	248.	163.	80.	34.	31.	67.	203.	328.	413.
DALLAS TX (LAT. 32.5)												
$\bar{H}$ MJ/m^2	9.67	12.85	16.49	19.01	21.81	24.91	24.62	22.52	19.17	15.20	10.93	9.25
$\bar{K}_T$	.49	.53	.54	.53	.55	.61	.61	.60	.59	.58	.53	.51
$\bar{T}_a$ C	7.0	10.0	13.0	19.0	23.0	28.0	30.0	30.0	26.0	20.0	13.0	9.0
DD C-DAY	338.	243.	174.	39.	0.	0.	0.	0.	0.	31.	158.	289.
DAVIS CA (LAT. 38.3)												
$\bar{H}$ MJ/m^2	6.61	10.71	16.81	22.08	26.60	29.36	28.86	25.55	20.83	14.55	9.03	6.19
$\bar{K}_T$	.41	.50	.60	.64	.67	.71	.72	.70	.69	.63	.52	.42
$\bar{T}_a$ C	7.0	9.0	11.0	14.0	17.0	21.0	23.0	22.0	21.0	17.0	11.0	7.0
DD C-DAY	324.	230.	184.	99.	40.	0.	0.	0.	0.	31.	178.	303.
DAYTON OH (LAT. 39.5)												
$\bar{H}$ MJ/m^2	6.78	9.38	13.90	17.46	21.69	24.07	23.61	21.52	17.75	12.94	7.83	6.07
$\bar{K}_T$	.44	.46	.51	.51	.55	.58	.59	.59	.59	.57	.47	.44
$\bar{T}_a$ C	-2.0	-1.0	4.0	11.0	16.0	22.0	24.0	23.0	19.0	13.0	5.0	-1.0
DD C-DAY	636.	538.	448.	229.	92.	7.	0.	4.	35.	171.	387.	587.
DENVER CO (LAT. 39.4)												
$\bar{H}$ MJ/m^2	10.68	14.15	18.25	21.73	24.37	27.38	26.50	24.79	20.68	15.49	10.97	9.13
$\bar{K}_T$	.69	.69	.67	.63	.62	.66	.66	.68	.69	.69	.66	.65
$\bar{T}_a$ C	-1.0	.0	3.0	9.0	14.0	19.0	23.0	22.0	17.0	11.0	4.0	.0
DD C-DAY	604.	501.	482.	292.	141.	44.	0.	0.	67.	227.	427.	558.
DES MOINES IA (LAT. 41.3)												
$\bar{H}$ MJ/m^2	7.03	9.92	13.48	17.79	21.52	23.78	23.78	20.93	16.96	12.60	7.91	5.78
$\bar{K}_T$	.49	.51	.51	.52	.55	.57	.59	.58	.58	.58	.51	.45
$\bar{T}_a$ C	-7.0	-4.0	1.0	10.0	16.0	21.0	24.0	23.0	18.0	12.0	3.0	-4.0
DD C-DAY	786.	634.	536.	258.	103.	14.	0.	7.	52.	194.	453.	689.

	JAN	FEB	MAR	APR	MAY	JUNE	JULY	AUG	SEP	OCT	NOV	DEC
DETROIT	MI (LAT. 42.1)											
$\overline{H}$ MJ/m^2	5.44	8.33	12.52	16.20	20.89	23.24	23.36	20.26	16.03	11.39	6.20	4.81
$\overline{K}_T$	.40	.44	.48	.48	.53	.56	.58	.57	.56	.54	.41	.39
$\overline{T}_a$ C	-4.0	-3.0	2.0	9.0	14.0	20.0	22.0	22.0	18.0	12.0	4.0	-2.0
DD C-DAY	696.	597.	512.	288.	136.	20.	3.	9.	53.	209.	415.	629.
DODGE CITY	KA (LAT. 37.5)											
$\overline{H}$ MJ/m2	10.83	13.67	18.07	22.58	23.54	27.56	27.18	24.38	20.62	15.89	11.71	9.70
$\overline{K}_T$	.65	.63	.64	.65	.60	.67	.67	.67	.67	.67	.66	.64
$\overline{T}_a$ C	-1.0	1.0	5.0	12.0	17.0	23.0	26.0	25.0	20.0	13.0	5.0	.0
DD C-DAY	589.	463.	410.	191.	64.	12.	0.	0.	23.	137.	370.	544.
DULUTH	MN (LAT. 46.5)											
$\overline{H}$ MJ/m^2	5.57	8.83	13.44	16.70	20.22	23.11	23.19	19.55	13.94	9.71	5.32	4.35
$\overline{K}_T$	.50	.53	.56	.52	.52	.56	.58	.56	.52	.52	.43	.45
$\overline{T}_a$ C	-13.0	-11.0	-5.0	4.0	10.0	15.0	19.0	18.0	12.0	7.0	-2.0	-10.0
DD C-DAY	973.	823.	715.	440.	269.	108.	37.	58.	177.	339.	610.	872.
EAST LANSING	MI (LAT. 42.4)											
$\overline{H}$ MJ/m^2	4.81	8.36	12.29	14.18	19.65	21.70	21.37	18.44	14.76	10.12	5.39	4.31
$\overline{K}_T$	.35	.44	.47	.42	.50	.52	.53	.52	.51	.48	.36	.35
$\overline{T}_a$ C	-5.0	-4.0	.0	8.0	13.0	19.0	21.0	20.0	16.0	10.0	3.0	-2.0
DD C-DAY	730.	638.	553.	308.	156.	27.	5.	15.	74.	234.	443.	653.
EL PASO	TX (LAT. 31.5)											
$\overline{H}$ MJ/m^2	13.84	18.07	22.96	27.39	29.90	30.53	28.02	26.72	24.05	19.32	15.35	13.09
$\overline{K}_T$	.69	.72	.75	.76	.75	.75	.70	.71	.74	.73	.72	.70
$\overline{T}_a$ C	7.0	9.0	13.0	17.0	22.0	27.0	27.0	26.0	23.0	18.0	11.0	7.0
DD C-DAY	381.	247.	177.	58.	0.	0.	0.	0.	0.	47.	230.	360.
ELY	NV (LAT. 39.2)											
$\overline{H}$ MJ/m^2	9.95	13.93	19.40	23.59	26.10	29.61	27.10	25.43	21.70	16.43	12.00	9.20
$\overline{K}_T$	.64	.67	.70	.68	.66	.72	.67	.70	.72	.72	.72	.65
$\overline{T}_a$ C	-4.0	-2.0	.0	5.0	10.0	14.0	19.0	18.0	13.0	7.0	1.0	-3.0
DD C-DAY	727.	597.	543.	373.	253.	125.	16.	24.	130.	329.	522.	658.

	JAN	FEB	MAR	APR	MAY	JUNE	JULY	AUG	SEP	OCT	NOV	DEC
FAIRBANKS	AK (LAT. 64.5)											
$\overline{H}$ MJ/m^2	.80	3.18	9.74	16.10	19.95	22.04	18.57	15.18	7.69	3.60	1.13	.25
$\overline{K}_T$	.61	.55	.70	.64	.56	.54	.49	.52	.43	.45	.51	.49
$\overline{T}_a$ C	-24.0	-19.0	-13.0	-1.0	8.0	14.0	15.0	12.0	6.0	-3.0	-16.0	-22.0
DD C-DAY	1311.	1056.	966.	593.	308.	123.	95.	184.	357.	668.	1018.	1252.
FARGO	ND (LAT. 46.5)											
$\overline{H}$ MJ/m^2	5.32	8.92	12.89	17.54	21.10	22.02	23.19	19.89	14.57	10.13	5.48	4.94
$\overline{K}_T$	.48	.54	.54	.54	.55	.53	.58	.57	.54	.54	.44	.51
$\overline{T}_a$ C	-14.0	-12.0	-4.0	6.0	13.0	18.0	21.0	21.0	14.0	8.0	-2.0	-11.0
DD C-DAY	1018.	844.	703.	378.	186.	54.	7.	18.	130.	310.	607.	896.
FORT_SMITH	AR (LAT. 35.2)											
$\overline{H}$ MJ/m^2	8.25	11.30	15.11	18.80	22.02	23.61	22.86	21.86	18.13	14.19	9.80	7.83
$\overline{K}_T$	.46	.49	.52	.53	.56	.57	.57	.59	.58	.57	.51	.47
$\overline{T}_a$ C	4.0	6.0	10.0	17.0	21.0	26.0	28.0	27.0	23.0	17.0	10.0	5.0
DD C-DAY	448.	338.	262.	73.	9.	0.	0.	0.	0.	75.	243.	405.
FORT_WAYNE	IN (LAT. 41.0)											
$\overline{H}$ MJ/m^2	6.32	9.08	13.44	17.08	22.06	24.45	23.82	21.10	16.54	12.35	7.07	5.65
$\overline{K}_T$	.44	.46	.50	.50	.56	.59	.59	.59	.56	.57	.45	.43
$\overline{T}_a$ C	-4.0	-2.0	2.0	10.0	15.0	21.0	23.0	22.0	18.0	12.0	5.0	-2.0
DD C-DAY	684.	582.	491.	262.	120.	13.	0.	7.	50.	202.	413.	627.
FORT_WORTH	TX (LAT. 32.5)											
$\overline{H}$ MJ/m^2	10.54	13.42	17.77	12.25	23.46	26.85	25.59	24.59	20.91	16.48	12.46	10.20
$\overline{K}_T$	.54	.55	.59	.34	.59	.66	.64	.66	.65	.63	.60	.56
$\overline{T}_a$ C	7.0	9.0	13.0	18.0	22.0	27.0	29.0	29.0	25.0	19.0	13.0	8.0
DD C-DAY	341.	249.	177.	55.	0.	0.	0.	0.	0.	36.	180.	298.
FRESNO	CA (LAT. 36.5)											
$\overline{H}$ MJ/m^2	7.78	12.38	18.32	22.79	26.64	29.15	27.93	25.34	21.03	15.68	10.08	6.69
$\overline{K}_T$	.45	.55	.64	.65	.67	.71	.69	.69	.68	.65	.55	.42
$\overline{T}_a$ C	7.0	10.0	12.0	16.0	19.0	23.0	27.0	26.0	23.0	18.0	12.0	7.0
DD C-DAY	336.	237.	186.	90.	34.	3.	0.	0.	0.	47.	197.	321.

	JAN	FEB	MAR	APR	MAY	JUNE	JULY	AUG	SEP	OCT	NOV	DEC
GAINESVILLE FL (LAT. 29.4)												
$\overline{H}$ MJ/m^2	11.63	15.35	18.61	22.54	24.51	22.75	21.75	21.24	18.57	15.39	13.30	10.62
$\overline{K}_T$	.54	.59	.59	.62	.62	.56	.54	.56	.56	.56	.59	.53
$\overline{T}_a$ C	13.0	14.0	17.0	21.0	24.0	26.0	27.0	27.0	26.0	22.0	17.0	14.0
DD C-DAY	164.	133.	73.	11.	0.	0.	0.	0.	0.	7.	69.	143.
GLASGOW MT (LAT. 48.1)												
$\overline{H}$ MJ/m^2	6.44	10.58	15.77	19.03	23.50	25.63	26.76	22.33	17.15	11.21	6.48	4.94
$\overline{K}_T$	.64	.68	.68	.60	.61	.62	.67	.65	.65	.63	.57	.57
$\overline{T}_a$ C	-12.0	-8.0	-3.0	6.0	12.0	17.0	21.0	21.0	14.0	8.0	-1.0	-7.0
DD C-DAY	951.	799.	659.	360.	186.	83.	17.	26.	150.	338.	613.	814.
GRAND JUNCTION CO (LAT. 39.1)												
$\overline{H}$ MJ/m^2	9.70	13.59	17.98	22.29	25.30	29.61	28.06	24.30	20.95	15.81	11.00	9.03
$\overline{K}_T$	.62	.65	.65	.65	.64	.72	.70	.67	.70	.69	.65	.64
$\overline{T}_a$ C	-3.0	.0	5.0	11.0	16.0	22.0	25.0	24.0	19.0	12.0	4.0	-1.0
DD C-DAY	672.	504.	405.	215.	81.	12.	0.	0.	17.	174.	437.	617.
GRAND LAKE CO (LAT. 40.2)												
$\overline{H}$ MJ/m^2	8.88	13.11	17.71	21.44	23.11	26.46	25.12	21.14	19.93	15.11	9.80	7.70
$\overline{K}_T$	.59	.65	.65	.63	.59	.64	.62	.58	.67	.68	.61	.57
$\overline{T}_a$ C	-9.0	-7.0	-4.0	.0	6.0	10.0	13.0	12.0	8.0	3.0	-3.0	-8.0
DD C-DAY	864.	734.	720.	525.	381.	250.	153.	174.	280.	446.	653.	820.
GREAT FALLS MT (LAT. 47.3)												
$\overline{H}$ MJ/m^2	5.77	9.58	15.14	17.94	21.91	24.71	26.56	22.12	16.89	10.96	6.44	4.68
$\overline{K}_T$	.54	.59	.64	.56	.57	.60	.67	.64	.63	.60	.54	.51
$\overline{T}_a$ C	-5.0	-2.0	.0	6.0	12.0	16.0	21.0	20.0	14.0	9.0	1.0	-2.0
DD C-DAY	749.	641.	591.	357.	213.	103.	16.	29.	143.	302.	512.	649.
GREEN BAY WI (LAT. 44.3)												
$\overline{H}$ MJ/m^2	5.74	8.79	13.10	16.08	20.47	22.69	22.52	19.34	14.78	10.05	5.82	4.60
$\overline{K}_T$	.46	.49	.52	.49	.53	.55	.56	.55	.53	.50	.42	.42
$\overline{T}_a$ C	-9.0	-8.0	-2.0	7.0	12.0	18.0	21.0	20.0	15.0	10.0	1.0	-6.0
DD C-DAY	854.	731.	627.	353.	188.	51.	12.	30.	106.	272.	515.	759.

	JAN	FEB	MAR	APR	MAY	JUNE	JULY	AUG	SEP	OCT	NOV	DEC
GREENSBORO NC (LAT. 36.0)												
$\bar{H}$ MJ/m^2	8.57	11.37	14.80	19.61	22.29	23.50	22.62	19.86	17.23	13.76	10.16	7.78
$\bar{K}_T$	.49	.50	.51	.56	.56	.57	.56	.54	.55	.56	.55	.48
$\bar{T}_a$ C	3.0	4.0	8.0	14.0	19.0	23.0	25.0	24.0	21.0	14.0	8.0	4.0
DD C-DAY	453.	379.	302.	113.	33.	0.	0.	0.	13.	116.	278.	437.
GRNVLE-SPTNBRG NC (LAT. 34.5)												
$\bar{H}$ MJ/m^2	9.38	11.30	16.29	21.39	23.15	23.32	23.19	21.48	17.75	14.95	10.76	8.58
$\bar{K}_T$	.51	.48	.55	.60	.58	.57	.58	.58	.56	.60	.55	.51
$\bar{T}_a$ C	6.0	7.0	10.0	16.0	21.0	24.0	26.0	25.0	22.0	16.0	11.0	6.0
DD C-DAY	391.	321.	250.	80.	16.	0.	0.	0.	5.	81.	233.	381.
GRIFFIN GA (LAT. 33.1)												
$\bar{H}$ MJ/m^2	9.95	12.63	16.23	21.70	24.13	24.25	23.38	21.87	18.27	15.56	12.04	8.78
$\bar{K}_T$	.52	.52	.54	.60	.61	.59	.58	.59	.57	.60	.59	.49
$\bar{T}_a$ C	6.0	8.0	11.0	16.0	21.0	24.0	25.0	25.0	22.0	17.0	11.0	7.0
DD C-DAY	356.	289.	228.	61.	11.	0.	0.	0.	11.	61.	200.	339.
HARTFORD CT (LAT. 41.6)												
$\bar{H}$ MJ/m^2	6.62	9.46	13.73	16.12	19.85	22.36	22.15	19.22	15.24	11.05	6.91	7.91
$\bar{K}_T$	.47	.49	.52	.48	.51	.54	.55	.54	.52	.51	.45	.62
$\bar{T}_a$ C	-4.0	-3.0	2.0	9.0	15.0	20.0	23.0	21.0	17.0	11.0	5.0	-2.0
DD C-DAY	692.	594.	506.	288.	126.	13.	0.	7.	59.	213.	395.	634.
HILO HI (LAT. 19.4)												
$\bar{H}$ MJ/m^2	11.71	15.47	18.73	18.02	18.73	24.00	22.33	20.49	18.90	14.18	12.63	11.00
$\bar{K}_T$	.43	.50	.54	.48	.48	.61	.58	.54	.53	.45	.45	.43
$\bar{T}_a$ C	21.0	21.0	21.0	22.0	22.0	23.0	23.0	24.0	24.0	23.0	22.0	21.0
DD C-DAY	0.	0.	0.	0.	0.	0.	0.	0.	0.	0.	0.	0.
HONOLULU HI (LAT. 21.2)												
$\bar{H}$ MJ/m^2	15.20	17.67	21.60	23.40	25.83	25.75	25.75	25.62	23.99	21.23	17.84	15.53
$\bar{K}_T$	.58	.59	.63	.62	.66	.65	.66	.68	.68	.68	.66	.63
$\bar{T}_a$ C	22.0	22.0	22.0	23.0	24.0	25.0	26.0	26.0	26.0	25.0	24.0	23.0
DD C-DAY	0.	0.	0.	0.	0.	0.	0.	0.	0.	0.	0.	0.

	JAN	FEB	MAR	APR	MAY	JUNE	JULY	AUG	SEP	OCT	NOV	DEC
HOUSTON	TX (LAT. 29.6)											
$\overline{H}$ MJ/m^2	10.05	12.64	16.37	18.97	23.11	25.08	24.53	21.69	19.09	16.58	11.55	9.50
$\overline{K}_T$	.47	.49	.52	.52	.58	.62	.61	.58	.58	.60	.52	.48
$\overline{T}_a$ C	11.0	13.0	16.0	21.0	24.0	27.0	28.0	29.0	26.0	22.0	16.0	13.0
DD C-DAY	231.	163.	105.	13.	0.	0.	0.	0.	0.	13.	86.	185.
INDIANAPOLIS	IN (LAT. 39.4)											
$\overline{H}$ MJ/m^2	6.15	8.95	13.05	16.43	20.53	22.87	22.67	20.32	16.94	12.25	7.36	5.44
$\overline{K}_T$	.40	.43	.48	.48	.52	.55	.56	.56	.57	.54	.44	.39
$\overline{T}_a$ C	-1.0	.0	4.0	11.0	17.0	22.0	24.0	23.0	19.0	13.0	5.0	.0
DD C-DAY	639.	533.	436.	215.	88.	6.	0.	3.	35.	168.	388.	587.
INYOKERN	CA (LAT. 35.4)											
$\overline{H}$ MJ/m^2	13.05	17.52	24.17	29.32	32.99	34.96	32.79	30.86	27.10	20.24	15.31	12.34
$\overline{K}_T$	.73	.77	.83	.83	.83	.85	.81	.84	.86	.82	.80	.75
$\overline{T}_a$ C	7.0	11.0	14.0	18.0	23.0	27.0	32.0	31.0	27.0	20.0	13.0	8.0
DD C-DAY	341.	218.	148.	71.	6.	0.	0.	0.	0.	24.	176.	334.
ITHACA	NY (LAT. 42.3)											
$\overline{H}$ MJ/m^2	5.10	8.49	11.79	14.55	19.61	22.54	22.37	19.24	14.89	10.37	5.23	4.14
$\overline{K}_T$	.37	.45	.45	.43	.50	.54	.56	.54	.52	.49	.35	.34
$\overline{T}_a$ C	-5.0	-4.0	.0	7.0	12.0	18.0	20.0	19.0	15.0	10.0	4.0	-2.0
DD C-DAY	723.	646.	562.	332.	176.	39.	11.	22.	87.	243.	423.	654.
JACKSON	MS (LAT. 32.2)											
$\overline{H}$ MJ/m^2	8.88	11.72	15.83	20.18	22.94	23.53	22.78	21.39	17.96	15.11	10.42	8.46
$\overline{K}_T$	.45	.48	.52	.56	.58	.57	.57	.57	.55	.58	.50	.46
$\overline{T}_a$ C	8.0	10.0	13.0	19.0	23.0	26.0	28.0	27.0	24.0	19.0	13.0	9.0
DD C-DAY	316.	246.	174.	41.	3.	0.	0.	0.	0.	51.	167.	280.
JACKSONVILLE	FL (LAT. 30.2)											
$\overline{H}$ MJ/m^2	11.18	14.49	17.71	21.52	23.28	21.98	21.86	19.93	16.03	13.86	11.47	9.63
$\overline{K}_T$	.53	.57	.57	.59	.59	.54	.55	.53	.49	.51	.52	.49
$\overline{T}_a$ C	12.0	13.0	16.0	20.0	23.0	26.0	27.0	27.0	25.0	21.0	16.0	12.0
DD C-DAY	193.	157.	98.	13.	0.	0.	0.	0.	0.	11.	89.	176.

	JAN	FEB	MAR	APR	MAY	JUNE	JULY	AUG	SEP	OCT	NOV	DEC
KANSAS CITY MO (LAT. 39.2)												
$\overline{H}$ MJ/m^2	7.62	10.55	14.28	18.46	21.81	24.70	24.24	22.02	17.79	13.40	9.00	6.87
$\overline{K}_T$	.49	.51	.52	.54	.55	.60	.60	.61	.59	.59	.54	.49
$\overline{T}_a$ C	-2.0	1.0	5.0	13.0	18.0	23.0	26.0	25.0	20.0	15.0	6.0	.0
DD C-DAY	641.	496.	414.	174.	62.	7.	0.	0.	23.	131.	357.	563.
KEY WEST FL (LAT. 24.3)												
$\overline{H}$ MJ/m^2	13.69	17.16	20.51	23.95	24.24	22.73	22.36	20.97	18.63	16.49	13.90	12.23
$\overline{K}_T$	.56	.60	.62	.64	.62	.57	.57	.55	.54	.55	.55	.53
$\overline{T}_a$ C	21.0	21.0	23.0	25.0	27.0	28.0	29.0	29.0	28.0	26.0	23.0	21.0
DD C-DAY	9.	14.	3.	0.	0.	0.	0.	0.	0.	0.	0.	10.
LAKE CHARLES LA (LAT. 30.1)												
$\overline{H}$ MJ/m^2	10.00	12.71	16.56	20.20	23.17	24.34	21.79	21.08	18.73	16.81	12.38	9.70
$\overline{K}_T$	.48	.50	.53	.55	.58	.60	.54	.56	.57	.62	.56	.49
$\overline{T}_a$ C	11.0	12.0	15.0	20.0	23.0	27.0	28.0	27.0	25.0	21.0	15.0	12.0
DD C-DAY	212.	152.	108.	22.	0.	0.	0.	0.	0.	11.	117.	189.
LANDER WY (LAT. 42.5)												
$\overline{H}$ MJ/m^2	9.62	13.42	18.86	23.13	24.46	28.23	27.10	24.25	19.40	14.89	9.91	8.24
$\overline{K}_T$	.71	.71	.72	.69	.63	.68	.67	.68	.68	.71	.67	.68
$\overline{T}_a$ C	-6.0	-3.0	.0	6.0	12.0	16.0	21.0	21.0	15.0	8.0	.0	-4.0
DD C-DAY	787.	636.	565.	363.	212.	85.	3.	11.	113.	308.	567.	722.
LANSING MI (LAT. 42.5)												
$\overline{H}$ MJ/m^2	5.65	8.92	12.89	14.99	20.89	23.15	22.78	20.14	15.78	10.80	5.86	4.69
$\overline{K}_T$	.42	.47	.50	.45	.53	.56	.57	.56	.55	.51	.40	.39
$\overline{T}_a$ C	-5.0	-4.0	1.0	8.0	14.0	19.0	22.0	21.0	16.0	11.0	4.0	-3.0
DD C-DAY	730.	638.	553.	308.	156.	27.	5.	15.	74.	234.	443.	653.
LARAMIE WY (LAT. 41.2)												
$\overline{H}$ MJ/m^2	9.37	12.46	17.73	20.83	22.92	26.22	24.80	22.00	17.56	13.34	9.49	7.65
$\overline{K}_T$	.65	.63	.67	.61	.58	.63	.62	.61	.60	.62	.61	.59
$\overline{T}_a$ C	-5.0	-4.0	-2.0	3.0	9.0	14.0	18.0	17.0	12.0	6.0	.0	-4.0
DD C-DAY	763.	650.	653.	453.	296.	143.	39.	56.	192.	374.	577.	715.

	JAN	FEB	MAR	APR	MAY	JUNE	JULY	AUG	SEP	OCT	NOV	DEC
LAS VEGAS NV (LAT. 36.0)												
$\overline{H}$ MJ/m^2	11.67	16.14	21.12	26.01	29.36	31.20	28.27	26.22	23.29	17.94	13.30	10.87
$\overline{K}_T$	.67	.72	.73	.74	.74	.76	.70	.71	.75	.74	.71	.68
$\overline{T}_a$ C	6.0	9.0	12.0	17.0	23.0	28.0	31.0	30.0	26.0	19.0	11.0	7.0
DD C-DAY	382.	271.	186.	62.	3.	0.	0.	0.	0.	43.	215.	343.
LEMONT IL (LAT. 41.4)												
$\overline{H}$ MJ/m^2	7.15	9.70	13.63	16.31	20.78	23.13	22.04	20.32	16.06	11.08	6.57	5.48
$\overline{K}_T$	.50	.50	.51	.48	.53	.56	.55	.57	.55	.51	.43	.43
$\overline{T}_a$ C	-3.0	-2.0	3.0	10.0	16.0	21.0	24.0	23.0	19.0	13.0	5.0	-1.0
DD C-DAY	701.	585.	486.	252.	116.	14.	0.	4.	32.	176.	410.	629.
LEXINGTON KY (LAT. 38.0)												
$\overline{H}$ MJ/m^2	7.15	10.79	15.22	19.91	24.09	25.97	25.76	23.34	20.53	15.10	9.83	7.15
$\overline{K}_T$	.44	.50	.54	.57	.61	.63	.64	.64	.67	.65	.56	.48
$\overline{T}_a$ C	.0	2.0	6.0	13.0	18.0	23.0	25.0	24.0	20.0	14.0	7.0	2.0
DD C-DAY	553.	462.	374.	168.	59.	4.	0.	0.	22.	137.	340.	508.
LINCOLN NE (LAT. 40.5)												
$\overline{H}$ MJ/m^2	7.95	10.66	14.51	17.73	20.74	22.79	22.46	21.24	17.23	13.59	8.66	7.19
$\overline{K}_T$	.54	.53	.54	.52	.53	.55	.56	.59	.58	.62	.54	.54
$\overline{T}_a$ C	-4.0	-2.0	3.0	11.0	16.0	22.0	25.0	24.0	19.0	12.0	4.0	-1.0
DD C-DAY	687.	564.	463.	223.	95.	17.	0.	3.	42.	167.	405.	592.
LITTLE ROCK AR (LAT. 34.4)												
$\overline{H}$ MJ/m^2	8.28	10.96	14.97	19.03	22.08	23.50	23.34	21.58	18.48	14.47	10.20	7.82
$\overline{K}_T$	.45	.47	.51	.53	.56	.57	.58	.58	.58	.57	.52	.46
$\overline{T}_a$ C	4.0	6.0	10.0	16.0	21.0	26.0	27.0	27.0	23.0	17.0	10.0	5.0
DD C-DAY	420.	321.	241.	70.	5.	0.	0.	0.	5.	71.	258.	398.
LOS ANGELES CA (LAT. 33.6)												
$\overline{H}$ MJ/m^2	10.75	14.39	19.19	21.66	24.09	25.80	27.14	24.63	21.08	15.39	12.25	10.29
$\overline{K}_T$	.57	.60	.64	.61	.61	.63	.67	.66	.66	.60	.61	.59
$\overline{T}_a$ C	12.0	13.0	13.0	15.0	16.0	18.0	20.0	20.0	20.0	18.0	15.0	13.0
DD C-DAY	184.	150.	148.	108.	63.	39.	11.	8.	13.	43.	88.	155.

	JAN	FEB	MAR	APR	MAY	JUNE	JULY	AUG	SEP	OCT	NOV	DEC
LOUISVILLE	KY (LAT. 38.1)											
$\bar{H}$ MJ/m^2	6.87	9.67	13.61	17.58	21.56	23.45	23.03	20.85	17.08	12.69	7.95	6.28
$\bar{K}_T$	.42	.45	.49	.51	.55	.57	.57	.57	.56	.54	.46	.42
$\bar{T}_a$ C	.0	2.0	6.0	13.0	18.0	22.0	24.0	24.0	20.0	14.0	7.0	1.0
DD C-DAY	546.	454.	367.	159.	58.	0.	0.	0.	19.	134.	333.	506.
LYNN	MA (LAT. 42.3)											
$\bar{H}$ MJ/m^2	4.94	8.74	12.55	16.48	18.99	22.50	22.54	17.65	14.22	9.66	5.56	4.18
$\bar{K}_T$	.36	.46	.48	.49	.49	.54	.56	.49	.49	.46	.37	.34
$\bar{T}_a$ C	-1.0	.0	3.0	9.0	15.0	20.0	23.0	22.0	18.0	13.0	7.0	1.0
DD C-DAY	604.	540.	470.	285.	116.	20.	0.	5.	33.	176.	335.	546.
MACON	GA (LAT. 32.4)											
$\bar{H}$ MJ/m^2	10.38	12.98	16.41	21.39	23.70	23.95	23.07	21.81	18.13	15.41	11.14	9.08
$\bar{K}_T$	.53	.53	.54	.59	.60	.58	.57	.58	.56	.59	.54	.50
$\bar{T}_a$ C	9.0	10.0	14.0	19.0	23.0	26.0	27.0	27.0	24.0	19.0	13.0	9.0
DD C-DAY	302.	235.	166.	37.	3.	0.	0.	0.	0.	46.	169.	288.
MADISON	WI (LAT. 43.1)											
$\bar{H}$ MJ/m^2	6.41	9.22	13.99	16.53	19.82	23.07	23.24	19.76	16.40	11.28	6.31	5.63
$\bar{K}_T$	.49	.50	.54	.49	.51	.56	.58	.56	.58	.55	.44	.48
$\bar{T}_a$ C	-7.0	-6.0	.0	7.0	13.0	19.0	21.0	20.0	15.0	10.0	1.0	-5.0
DD C-DAY	830.	696.	599.	328.	165.	40.	8.	22.	96.	263.	505.	742.
MANHATTAN	KA (LAT. 39.1)											
$\bar{H}$ MJ/m^2	8.04	11.05	14.44	18.13	22.06	23.07	22.23	22.02	17.17	12.23	9.50	6.53
$\bar{K}_T$	.52	.53	.52	.53	.56	.56	.55	.61	.57	.54	.57	.46
$\bar{T}_a$ C	-1.0	1.0	5.0	13.0	18.0	23.0	26.0	25.0	20.0	14.0	6.0	.0
DD C-DAY	623.	496.	401.	183.	69.	7.	0.	0.	32.	150.	373.	544.
MATANUSKA	AK (LAT. 61.3)											
$\bar{H}$ MJ/m^2	1.34	3.85	10.13	14.90	18.25	19.34	17.12	13.15	8.29	4.18	1.59	.63
$\bar{K}_T$	.49	.50	.64	.56	.51	.48	.45	.43	.42	.42	.42	.37
$\bar{T}_a$ C	-11.0	-7.0	-3.0	2.0	8.0	12.0	14.0	12.0	8.0	1.0	-6.0	-10.0
DD C-DAY	914.	714.	689.	477.	310.	168.	129.	169.	288.	526.	738.	904.

		JAN	FEB	MAR	APR	MAY	JUNE	JULY	AUG	SEP	OCT	NOV	DEC
MEDFORD		OR (LAT. 42.2)											
$\overline{H}$	MJ/m^2	4.94	8.87	13.88	20.24	24.63	27.31	29.23	25.17	18.78	11.63	6.36	3.85
$\overline{K}_T$		.36	.46	.53	.60	.63	.66	.73	.70	.65	.55	.43	.31
$\overline{T}_a$	C	3.0	5.0	7.0	10.0	14.0	18.0	22.0	21.0	18.0	12.0	6.0	3.0
DD	C-DAY	489.	369.	348.	247.	139.	52.	6.	12.	49.	200.	358.	470.
MEMPHIS		TN (LAT. 35.0)											
$\overline{H}$	MJ/m^2	8.04	11.18	15.03	19.68	23.19	24.66	24.41	22.40	18.50	14.82	9.96	7.70
$\overline{K}_T$		.44	.48	.51	.55	.59	.60	.61	.61	.59	.60	.52	.46
$\overline{T}_a$	C	5.0	7.0	11.0	17.0	22.0	26.0	28.0	27.0	23.0	17.0	10.0	6.0
DD	C-DAY	422.	330.	254.	73.	12.	0.	0.	0.	4.	79.	235.	384.
MIAMI		FL (LAT. 25.5)											
$\overline{H}$	MJ/m^2	14.34	17.40	20.53	22.75	23.08	22.21	22.46	21.24	18.69	16.27	14.80	13.34
$\overline{K}_T$		.61	.62	.63	.61	.59	.55	.57	.56	.55	.56	.60	.60
$\overline{T}_a$	C	19.0	19.0	21.0	23.0	25.0	27.0	27.0	28.0	27.0	25.0	22.0	20.0
DD	C-DAY	41.	31.	11.	0.	0.	0.	0.	0.	0.	0.	0.	36.
MIDLAND		TX (LAT. 31.6)											
$\overline{H}$	MJ/m^2	11.75	15.05	19.95	23.04	25.80	25.72	25.68	24.55	21.37	16.69	13.47	11.37
$\overline{K}_T$		.58	.60	.65	.64	.65	.63	.64	.66	.66	.63	.63	.61
$\overline{T}_a$	C	6.0	8.0	12.0	17.0	22.0	26.0	27.0	27.0	23.0	18.0	11.0	7.0
DD	C-DAY	362.	260.	179.	50.	0.	0.	0.	0.	0.	42.	212.	329.
MILWAUKEE		WI (LAT. 42.6)											
$\overline{H}$	MJ/m^2	6.24	8.79	13.10	16.83	21.27	23.65	23.57	20.31	16.41	11.18	6.74	5.02
$\overline{K}_T$		.46	.47	.50	.50	.54	.57	.59	.57	.57	.53	.46	.42
$\overline{T}_a$	C	-7.0	-5.0	.0	7.0	12.0	18.0	21.0	21.0	16.0	11.0	2.0	-4.0
DD	C-DAY	786.	661.	579.	338.	193.	50.	8.	20.	78.	244.	475.	703.
MINN-ST. PAUL		MN (LAT. 44.5)											
$\overline{H}$	MJ/m^2	5.99	9.21	12.81	16.54	20.18	22.44	22.82	19.51	14.99	10.80	6.03	4.69
$\overline{K}_T$		.49	.52	.51	.50	.52	.54	.57	.55	.54	.54	.44	.43
$\overline{T}_a$	C	-11.0	-9.0	-2.0	7.0	14.0	19.0	22.0	21.0	16.0	10.0	.0	-7.0
DD	C-DAY	909.	754.	632.	332.	151.	36.	6.	12.	96.	262.	543.	799.

		JAN	FEB	MAR	APR	MAY	JUNE	JULY	AUG	SEP	OCT	NOV	DEC
MT WEATHER	VA (LAT. 39.0)												
$\overline{H}$	MJ/m^2	7.20	11.47	14.15	17.33	21.27	21.98	21.35	18.00	15.70	11.76	8.46	7.03
$\overline{K}_T$		.46	.55	.51	.50	.54	.53	.53	.50	.52	.52	.50	.49
$\overline{T}_a$	C	-1.0	.0	3.0	10.0	15.0	20.0	22.0	21.0	18.0	12.0	5.0	.0
DD	C-DAY	615.	535.	453.	251.	102.	13.	0.	3.	44.	189.	370.	573.
NASHVILLE	TN (LAT. 36.1)												
$\overline{H}$	MJ/m^2	6.82	10.04	13.76	18.82	21.62	23.71	23.13	20.66	17.90	13.67	9.07	6.73
$\overline{K}_T$		.39	.45	.48	.53	.55	.58	.57	.56	.57	.56	.49	.42
$\overline{T}_a$	C	3.0	5.0	9.0	15.0	20.0	24.0	26.0	25.0	22.0	16.0	9.0	5.0
DD	C-DAY	460.	373.	291.	98.	25.	0.	0.	0.	6.	100.	277.	424.
NATICK	MA (LAT. 42.2)												
$\overline{H}$	MJ/m^2	6.36	9.70	13.63	16.35	20.91	17.10	21.45	19.11	15.01	10.91	6.15	5.60
$\overline{K}_T$		.46	.51	.52	.49	.53	.41	.53	.53	.52	.52	.41	.46
$\overline{T}_a$	C	-2.0	-2.0	3.0	9.0	15.0	20.0	23.0	22.0	17.0	12.0	6.0	.0
DD	C-DAY	672.	585.	487.	282.	123.	17.	0.	6.	56.	203.	375.	608.
NEW ORLEANS	LA (LAT. 29.6)												
$\overline{H}$	MJ/m^2	8.96	10.84	14.03	17.25	18.80	18.55	17.46	17.42	16.03	14.95	11.64	8.29
$\overline{K}_T$		.42	.42	.45	.47	.47	.46	.44	.46	.48	.54	.52	.42
$\overline{T}_a$	C	11.0	13.0	15.0	20.0	23.0	26.0	27.0	27.0	25.0	20.0	15.0	12.0
DD	C-DAY	202.	143.	107.	22.	0.	0.	0.	0.	0.	11.	107.	179.
NEWPORT	RI (LAT. 41.3)												
$\overline{H}$	MJ/m^2	6.48	9.66	13.80	16.52	20.45	22.50	21.62	18.78	15.89	11.42	7.32	5.90
$\overline{K}_T$		.45	.49	.52	.49	.52	.54	.54	.52	.54	.53	.47	.46
$\overline{T}_a$	C	.0	.0	3.0	7.0	12.0	17.0	21.0	21.0	18.0	13.0	8.0	2.0
DD	C-DAY	567.	531.	487.	340.	191.	55.	0.	9.	43.	171.	330.	501.
NEW YORK	NY (LAT. 40.5)												
$\overline{H}$	MJ/m^2	5.44	8.33	12.14	15.45	18.09	19.68	19.22	16.29	13.86	10.13	6.15	4.81
$\overline{K}_T$		.37	.41	.45	.45	.46	.48	.48	.45	.47	.46	.38	.36
$\overline{T}_a$	C	.0	1.0	5.0	11.0	17.0	22.0	25.0	24.0	20.0	15.0	9.0	2.0
DD	C-DAY	541.	488.	417.	230.	69.	3.	0.	0.	15.	124.	293.	493.

		JAN	FEB	MAR	APR	MAY	JUNE	JULY	AUG	SEP	OCT	NOV	DEC
NORFOLK	VA (LAT. 36.5)												
$\overline{H}$	MJ/m^2	8.71	11.30	15.57	19.97	22.61	23.95	23.03	20.14	16.66	12.98	9.34	7.70
$\overline{K}_T$		.51	.51	.54	.57	.57	.58	.57	.55	.54	.54	.51	.49
$\overline{T}_a$	C	4.0	5.0	8.0	14.0	19.0	23.0	25.0	24.0	22.0	16.0	10.0	5.0
DD	C-DAY	422.	367.	296.	126.	29.	0.	0.	0.	5.	78.	223.	391.
NORTH OMAHA	NE (LAT. 41.2)												
$\overline{H}$	MJ/m^2	8.55	11.60	14.90	19.38	21.48	23.57	23.78	21.81	16.58	12.31	8.29	6.95
$\overline{K}_T$		.60	.59	.56	.57	.55	.57	.59	.61	.57	.57	.53	.54
$\overline{T}_a$	C	-5.0	-3.0	3.0	10.0	17.0	22.0	25.0	23.0	19.0	12.0	4.0	-2.0
DD	C-DAY	753.	626.	522.	258.	116.	23.	0.	7.	58.	198.	460.	653.
OAK RIDGE	TN (LAT. 36.0)												
$\overline{H}$	MJ/m^2	6.94	9.95	13.63	18.69	21.54	22.79	21.79	19.91	17.44	13.34	8.82	6.73
$\overline{K}_T$		.40	.44	.47	.53	.54	.55	.54	.54	.56	.55	.47	.42
$\overline{T}_a$	C	3.0	5.0	9.0	15.0	19.0	23.0	25.0	24.0	21.0	15.0	8.0	4.0
DD	C-DAY	463.	383.	306.	122.	43.	0.	0.	0.	11.	120.	298.	444.
OKLAHOMA CITY	OK (LAT. 35.2)												
$\overline{H}$	MJ/m^2	10.66	13.26	17.02	20.83	22.58	26.35	25.51	24.59	20.24	15.85	11.88	9.91
$\overline{K}_T$		.59	.58	.58	.59	.57	.64	.63	.67	.64	.64	.62	.60
$\overline{T}_a$	C	2.0	4.0	9.0	15.0	20.0	25.0	27.0	27.0	23.0	16.0	9.0	4.0
DD	C-DAY	486.	369.	296.	100.	20.	0.	0.	0.	7.	82.	263.	431.
PAGE	AZ (LAT. 36.4)												
$\overline{H}$	MJ/m^2	12.56	15.99	22.02	25.87	29.10	29.60	28.47	24.95	21.60	16.83	12.98	10.17
$\overline{K}_T$		.73	.72	.77	.74	.74	.72	.71	.68	.70	.70	.70	.64
$\overline{T}_a$	C	.0	2.0	5.0	10.0	15.0	20.0	24.0	22.0	18.0	12.0	5.0	.0
DD	C-DAY	591.	447.	396.	240.	107.	21.	0.	6.	41.	189.	390.	562.
PARKERSBURG	WV (LAT. 39.2)												
$\overline{H}$	MJ/m^2	5.99	8.46	12.64	15.87	20.39	22.65	22.02	20.26	16.50	11.89	6.99	5.53
$\overline{K}_T$		.38	.41	.46	.46	.52	.55	.55	.56	.55	.52	.42	.39
$\overline{T}_a$	C	1.0	1.0	6.0	13.0	18.0	22.0	24.0	23.0	20.0	14.0	7.0	2.0
DD	C-DAY	553.	471.	381.	178.	67.	4.	0.	0.	26.	149.	333.	515.

	JAN	FEB	MAR	APR	MAY	JUNE	JULY	AUG	SEP	OCT	NOV	DEC
PASADENA CA (LAT. 34.1)												
$\bar{H}$ MJ/m^2	10.51	13.94	18.38	21.31	23.82	24.28	26.54	25.08	20.18	15.32	11.35	9.88
$\bar{K}_T$	.56	.59	.62	.60	.60	.59	.66	.68	.63	.60	.57	.57
$\bar{T}_a$ C	12.0	13.0	13.0	15.0	17.0	19.0	23.0	23.0	22.0	19.0	15.0	13.0
DD C-DAY	191.	151.	141.	94.	47.	26.	0.	0.	6.	29.	91.	166.
PENSACOLA FL (LAT. 30.3)												
$\bar{H}$ MJ/m^2	10.47	13.44	16.96	21.31	23.53	23.78	22.48	21.31	18.00	16.49	11.64	9.38
$\bar{K}_T$	.50	.53	.54	.59	.59	.58	.56	.57	.55	.61	.53	.48
$\bar{T}_a$ C	11.0	12.0	15.0	20.0	23.0	26.0	27.0	27.0	25.0	21.0	15.0	12.0
DD C-DAY	237.	179.	117.	21.	0.	0.	0.	0.	0.	18.	105.	199.
PEORIA IL (LAT. 40.4)												
$\bar{H}$ MJ/m^2	6.82	9.54	13.48	17.67	21.31	23.99	23.57	21.02	17.04	12.52	7.75	5.82
$\bar{K}_T$	.46	.47	.50	.52	.54	.58	.59	.58	.58	.57	.48	.43
$\bar{T}_a$ C	-5.0	-2.0	3.0	11.0	16.0	22.0	24.0	23.0	19.0	13.0	4.0	-2.0
DD C-DAY	709.	580.	477.	231.	100.	9.	0.	4.	39.	182.	418.	637.
PHOENIX AZ (LAT. 33.3)												
$\bar{H}$ MJ/m^2	12.42	17.06	21.79	26.89	30.28	30.95	27.27	25.59	23.75	18.90	14.18	11.71
$\bar{K}_T$	.65	.71	.73	.75	.76	.75	.68	.69	.74	.73	.70	.66
$\bar{T}_a$ C	10.0	13.0	15.0	19.0	24.0	29.0	32.0	31.0	28.0	22.0	15.0	11.0
DD C-DAY	238.	162.	103.	33.	0.	0.	0.	0.	0.	9.	101.	216.
PHILADELPHIA PA (LAT. 39.5)												
$\bar{H}$ MJ/m^2	7.33	10.13	14.53	17.79	20.64	23.19	22.52	19.47	16.24	12.27	8.00	6.36
$\bar{K}_T$	.48	.49	.53	.52	.52	.56	.56	.54	.54	.54	.48	.46
$\bar{T}_a$ C	.0	1.0	5.0	11.0	17.0	22.0	24.0	23.0	20.0	14.0	8.0	2.0
DD C-DAY	563.	484.	398.	204.	68.	0.	0.	0.	21.	138.	313.	513.
PITTSBURGH PA (LAT. 40.3)												
$\bar{H}$ MJ/m^2	6.62	8.92	13.48	16.75	20.39	23.40	22.90	20.18	17.00	12.31	7.70	5.94
$\bar{K}_T$	.44	.44	.50	.49	.52	.57	.57	.56	.57	.56	.48	.44
$\bar{T}_a$ C	.0	.0	5.0	11.0	17.0	22.0	24.0	23.0	19.0	13.0	7.0	1.0
DD C-DAY	592.	513.	424.	212.	89.	6.	0.	3.	32.	166.	348.	546.

	JAN	FEB	MAR	APR	MAY	JUNE	JULY	AUG	SEP	OCT	NOV	DEC
POCATELLO ID (LAT. 42.5)												
$\bar{H}$ MJ/m^2	6.91	10.47	15.28	21.77	24.28	27.42	28.26	24.70	19.89	13.82	8.58	6.49
$\bar{K}_T$	.51	.55	.59	.65	.62	.66	.70	.69	.69	.66	.58	.54
$\bar{T}_a$ C	-4.0	.0	2.0	7.0	12.0	17.0	22.0	21.0	15.0	9.0	2.0	-2.0
DD C-DAY	720.	554.	510.	328.	187.	77.	0.	11.	107.	286.	488.	656.
PORT ARTHUR TX (LAT. 29.6)												
$\bar{H}$ MJ/m^2	9.59	12.35	16.08	18.21	22.44	23.86	22.11	20.47	18.09	15.99	10.76	8.92
$\bar{K}_T$	.45	.48	.51	.50	.57	.59	.55	.54	.55	.58	.48	.45
$\bar{T}_a$ C	11.0	13.0	16.0	20.0	24.0	27.0	28.0	28.0	26.0	21.0	16.0	12.0
DD C-DAY	233.	168.	112.	18.	0.	0.	0.	0.	0.	19.	102.	190.
PORTLAND ME (LAT. 43.4)												
$\bar{H}$ MJ/m^2	6.57	9.91	15.01	16.98	21.45	22.62	23.46	20.16	16.02	11.42	6.57	5.77
$\bar{K}_T$	.51	.54	.59	.51	.55	.55	.58	.57	.56	.56	.46	.50
$\bar{T}_a$ C	-5.0	-4.0	.0	6.0	11.0	17.0	20.0	19.0	15.0	9.0	4.0	-3.0
DD C-DAY	744.	657.	579.	375.	207.	62.	7.	29.	108.	282.	448.	564.
PORTLAND OR (LAT. 45.4)												
$\bar{H}$ MJ/m^2	4.02	6.57	10.05	14.78	17.71	19.80	23.15	18.59	14.36	8.41	4.86	3.47
$\bar{K}_T$	.34	.38	.41	.45	.46	.48	.58	.53	.52	.43	.37	.34
$\bar{T}_a$ C	3.0	6.0	8.0	10.0	14.0	17.0	19.0	19.0	17.0	12.0	7.0	5.0
DD C-DAY	463.	346.	332.	240.	147.	71.	27.	31.	66.	193.	328.	418.
PROSSER WA (LAT. 46.1)												
$\bar{H}$ MJ/m^2	4.90	9.29	14.70	21.81	25.79	28.47	29.60	25.29	19.18	11.47	5.69	4.19
$\bar{K}_T$	.43	.55	.61	.67	.67	.69	.74	.73	.71	.61	.45	.42
$\bar{T}_a$ C	-1.0	4.0	6.0	10.0	14.0	18.0	21.0	20.0	16.0	10.0	4.0	.0
DD C-DAY	624.	428.	376.	240.	127.	47.	7.	16.	66.	231.	413.	543.
PUEBLO CO (LAT. 38.2)												
$\bar{H}$ MJ/m^2	11.39	14.74	18.67	22.78	25.20	28.05	27.13	25.08	21.10	16.50	12.31	10.01
$\bar{K}_T$	.70	.69	.67	.66	.64	.68	.67	.69	.69	.71	.71	.68
$\bar{T}_a$ C	-1.0	1.0	4.0	11.0	16.0	21.0	25.0	24.0	19.0	12.0	5.0	1.0
DD C-DAY	601.	471.	431.	225.	82.	16.	0.	0.	31.	186.	403.	551.

		JAN	FEB	MAR	APR	MAY	JUNE	JULY	AUG	SEP	OCT	NOV	DEC
PULLMAN	WA (LAT. 46.4)												
$\overline{H}$	MJ/m^2	5.14	7.61	12.42	19.07	22.67	28.69	29.52	23.08	17.90	10.71	6.15	4.01
$\overline{K}_T$		.46	.46	.52	.59	.59	.69	.74	.66	.66	.57	.50	.41
$\overline{T}_a$	C	-2.0	1.0	3.0	7.0	11.0	15.0	19.0	18.0	15.0	9.0	3.0	.0
DD	C-DAY	637.	481.	460.	317.	204.	105.	26.	42.	114.	271.	452.	572.
PUT-IN-BAY	OH (LAT. 41.4)												
$\overline{H}$	MJ/m^2	5.02	8.32	12.21	15.43	20.66	22.67	23.71	21.58	16.69	12.34	6.57	4.64
$\overline{K}_T$		.35	.43	.46	.46	.53	.55	.59	.60	.57	.57	.43	.36
$\overline{T}_a$	C	-2.0	-1.0	2.0	9.0	15.0	21.0	24.0	23.0	19.0	14.0	6.0	.0
DD	C-DAY	661.	576.	501.	283.	122.	13.	0.	0.	22.	154.	367.	589.
RALEIGH	NC (LAT. 35.5)												
$\overline{H}$	MJ/m^2	9.95	12.75	16.81	19.78	20.87	23.84	22.71	20.20	16.14	13.05	10.00	8.45
$\overline{K}_T$		.56	.56	.58	.56	.53	.58	.56	.55	.52	.53	.53	.52
$\overline{T}_a$	C	5.0	6.0	10.0	15.0	20.0	24.0	25.0	25.0	22.0	16.0	10.0	5.0
DD	C-DAY	422.	354.	279.	100.	27.	0.	0.	0.	7.	103.	250.	410.
RALEIGH-DURHAM	NC (LAT. 35.5)												
$\overline{H}$	MJ/m^2	9.00	11.85	15.66	20.22	21.64	22.44	23.15	20.14	16.62	13.23	10.05	8.16
$\overline{K}_T$		.51	.52	.54	.57	.55	.54	.57	.55	.53	.54	.53	.50
$\overline{T}_a$	C	5.0	6.0	10.0	15.0	20.0	24.0	25.0	25.0	21.0	16.0	10.0	5.0
DD	C-DAY	422.	354.	279.	100.	27.	0.	0.	0.	7.	103.	250.	410.
RAPID CTY	SD (LAT. 44.1)												
$\overline{H}$	MJ/m^2	7.78	11.63	16.69	20.28	22.41	24.76	24.88	22.62	17.98	13.13	8.57	6.61
$\overline{K}_T$		.62	.65	.66	.61	.58	.60	.62	.64	.64	.65	.62	.59
$\overline{T}_a$	C	-5.0	-2.0	.0	7.0	13.0	18.0	23.0	22.0	16.0	10.0	2.0	-2.0
DD	C-DAY	741.	636.	584.	342.	181.	70.	12.	7.	92.	267.	498.	651.
RENO	NV (LAT. 39.3)												
$\overline{H}$	MJ/m^2	9.80	13.56	18.80	24.79	27.80	29.89	29.60	27.05	22.27	16.54	11.60	8.75
$\overline{K}_T$		.63	.65	.68	.72	.71	.72	.73	.74	.74	.73	.69	.62
$\overline{T}_a$	C	.0	2.0	4.0	8.0	12.0	16.0	20.0	19.0	15.0	10.0	4.0	.0
DD	C-DAY	570.	434.	426.	303.	182.	81.	9.	28.	93.	253.	415.	551.

	JAN	FEB	MAR	APR	MAY	JUNE	JULY	AUG	SEP	OCT	NOV	DEC
RICHLAND	WA (LAT. 46.2)											
$\overline{H}$ MJ/m^2	3.60	8.41	13.93	19.53	21.58	27.06	24.09	25.13	16.23	9.58	5.19	4.14
$\overline{K}_T$	.32	.50	.57	.60	.56	.65	.60	.72	.60	.51	.41	.42
$\overline{T}_a$ C	.0	4.0	8.0	12.0	17.0	20.0	24.0	22.0	18.0	12.0	6.0	2.0
DD C-DAY	571.	395.	327.	183.	72.	17.	7.	16.	43.	202.	383.	503.
RICHMOND	VA (LAT. 37.5)											
$\overline{H}$ MJ/m^2	7.91	10.72	15.07	19.34	21.86	23.49	23.40	20.14	16.54	12.64	8.67	7.12
$\overline{K}_T$	.48	.49	.53	.55	.55	.57	.58	.55	.54	.54	.49	.47
$\overline{T}_a$ C	3.0	4.0	8.0	14.0	19.0	23.0	25.0	25.0	21.0	15.0	9.0	4.0
DD C-DAY	474.	398.	316.	126.	36.	0.	0.	0.	12.	113.	267.	448.
RIVERSIDE	CA (LAT. 33.6)											
$\overline{H}$ MJ/m^2	11.51	15.37	20.01	22.65	26.08	28.47	28.18	25.87	22.40	17.04	13.36	11.30
$\overline{K}_T$	.61	.64	.67	.63	.66	.69	.70	.70	.70	.67	.67	.64
$\overline{T}_a$ C	11.0	12.0	13.0	15.0	18.0	20.0	24.0	24.0	22.0	18.0	14.0	11.0
DD C-DAY	226.	173.	157.	93.	41.	12.	0.	0.	3.	34.	118.	208.
ROCHESTER	NY (LAT. 43.1)											
$\overline{H}$ MJ/m^2	5.65	8.41	12.64	16.66	21.52	23.95	24.03	20.51	15.53	10.55	6.03	4.77
$\overline{K}_T$	.43	.45	.49	.50	.55	.58	.60	.58	.55	.51	.42	.41
$\overline{T}_a$ C	-4.0	-4.0	1.0	8.0	14.0	19.0	22.0	21.0	17.0	11.0	5.0	-2.0
DD C-DAY	706.	626.	551.	315.	158.	26.	5.	14.	70.	221.	408.	632.
SACRAMENTO	CA (LAT. 38.3)											
$\overline{H}$ MJ/m^2	6.82	11.10	16.62	22.23	28.39	28.68	28.47	24.37	20.43	15.28	9.42	6.36
$\overline{K}_T$	.42	.52	.60	.64	.72	.69	.71	.67	.67	.66	.54	.43
$\overline{T}_a$ C	7.0	10.0	12.0	15.0	18.0	21.0	24.0	23.0	22.0	17.0	12.0	8.0
DD C-DAY	343.	237.	207.	126.	67.	11.	0.	0.	3.	56.	200.	331.
ST. CLOUD	MN (LAT. 45.3)											
$\overline{H}$ MJ/m^2	7.11	10.50	15.31	17.69	20.87	22.62	23.21	20.53	15.05	10.08	6.11	5.14
$\overline{K}_T$	.60	.61	.62	.54	.54	.55	.58	.59	.55	.52	.47	.50
$\overline{T}_a$ C	-12.0	-9.0	-2.0	6.0	13.0	18.0	21.0	20.0	14.0	9.0	.0	-8.0
DD C-DAY	966.	804.	673.	368.	180.	47.	10.	21.	127.	299.	583.	847.

	JAN	FEB	MAR	APR	MAY	JUNE	JULY	AUG	SEP	OCT	NOV	DEC
ST. LOUIS MO (LAT. 38.4)												
$\bar{H}$ MJ/m^2	7.37	10.30	14.49	18.00	22.15	23.95	23.78	20.93	17.50	12.89	8.71	6.45
$\bar{K}_T$	.46	.49	.52	.52	.56	.58	.59	.57	.58	.56	.51	.44
$\bar{T}_a$ C	.0	2.0	6.0	14.0	19.0	24.0	26.0	25.0	21.0	15.0	7.0	1.0
DD C-DAY	581.	465.	379.	151.	57.	6.	0.	0.	19.	124.	333.	523.
SALT LAKE CITY UT (LAT. 40.5)												
$\bar{H}$ MJ/m^2	6.82	10.72	14.82	20.06	23.87	26.00	25.96	23.07	18.67	13.23	8.54	6.11
$\bar{K}_T$	.46	.53	.55	.59	.61	.63	.64	.64	.63	.60	.53	.46
$\bar{T}_a$ C	-1.0	1.0	4.0	10.0	15.0	19.0	25.0	24.0	18.0	11.0	4.0	.0
DD C-DAY	637.	492.	437.	263.	132.	49.	0.	3.	58.	223.	432.	598.
SAN ANTONIO TX (LAT. 29.3)												
$\bar{H}$ MJ/m^2	11.58	14.51	17.52	18.82	22.54	25.26	26.14	24.34	20.49	16.52	12.17	10.54
$\bar{K}_T$	.54	.56	.56	.51	.57	.62	.65	.65	.62	.60	.54	.52
$\bar{T}_a$ C	11.0	12.0	16.0	20.0	24.0	27.0	28.0	28.0	26.0	21.0	15.0	12.0
DD C-DAY	238.	159.	108.	22.	0.	0.	0.	0.	0.	17.	113.	202.
SAN DIEGO CA (LAT. 32.4)												
$\bar{H}$ MJ/m^2	11.10	14.36	17.92	19.43	20.64	21.35	22.90	20.89	18.67	15.11	11.89	10.26
$\bar{K}_T$	.57	.59	.59	.54	.52	.52	.57	.56	.58	.58	.57	.56
$\bar{T}_a$ C	12.0	13.0	14.0	15.0	17.0	18.0	20.0	21.0	21.0	18.0	15.0	13.0
DD C-DAY	174.	132.	122.	80.	44.	29.	3.	0.	9.	24.	78.	143.
SAN FRANCISCO CA (LAT. 37.5)												
$\bar{H}$ MJ/m^2	8.16	11.85	17.08	21.44	24.20	24.99	22.61	20.01	17.75	13.90	9.63	7.33
$\bar{K}_T$	.49	.54	.60	.61	.61	.61	.56	.55	.58	.59	.54	.48
$\bar{T}_a$ C	10.0	12.0	12.0	13.0	14.0	15.0	15.0	15.0	17.0	16.0	14.0	11.0
DD C-DAY	243.	181.	184.	162.	143.	108.	112.	98.	57.	71.	129.	224.
SANTA MARIA CA (LAT. 34.5)												
$\bar{H}$ MJ/m^2	11.08	14.64	20.32	23.42	26.64	29.11	28.48	25.59	21.91	17.48	13.01	10.58
$\bar{K}_T$	.60	.63	.69	.66	.67	.71	.71	.69	.69	.70	.67	.62
$\bar{T}_a$ C	10.0	11.0	11.0	12.0	13.0	15.0	16.0	16.0	17.0	15.0	13.0	10.0
DD C-DAY	255.	206.	202.	157.	129.	92.	55.	52.	53.	81.	150.	217.

	JAN	FEB	MAR	APR	MAY	JUNE	JULY	AUG	SEP	OCT	NOV	DEC
SAVANNAH GA (LAT. 32.1)												
$\bar{H}$ MJ/m^2	10.30	13.15	16.87	21.64	23.57	23.19	22.44	20.97	16.87	14.57	11.10	8.96
$\bar{K}_T$	.52	.53	.55	.60	.59	.57	.56	.56	.52	.55	.53	.49
$\bar{T}_a$ C	10.0	11.0	14.0	19.0	23.0	26.0	27.0	27.0	25.0	20.0	14.0	10.0
DD C-DAY	268.	211.	142.	35.	0.	0.	0.	0.	0.	33.	141.	254.
SAULT ST. MARIE MI (LAT. 46.3)												
$\bar{H}$ MJ/m^2	5.56	9.45	14.89	17.52	22.00	23.00	23.96	19.95	13.47	9.03	4.39	3.97
$\bar{K}_T$	.49	.57	.62	.54	.57	.56	.60	.57	.50	.48	.35	.41
$\bar{T}_a$ C	-9.0	-10.0	-4.0	3.0	9.0	14.0	17.0	17.0	13.0	7.0	.0	-6.0
DD C-DAY	847.	767.	709.	450.	265.	112.	53.	58.	155.	322.	528.	759.
SCHENECTADY NY (LAT. 42.5)												
$\bar{H}$ MJ/m^2	5.44	8.41	11.46	14.22	17.31	18.78	18.57	16.69	12.55	9.16	5.39	4.35
$\bar{K}_T$	.40	.44	.44	.42	.44	.45	.46	.47	.44	.44	.36	.36
$\bar{T}_a$ C	-5.0	-4.0	1.0	8.0	15.0	20.0	23.0	21.0	17.0	11.0	4.0	-2.0
DD C-DAY	744.	641.	543.	302.	136.	20.	4.	11.	76.	234.	420.	656.
SEATTLE WA (LAT. 47.3)												
$\bar{H}$ MJ/m^2	3.26	5.69	11.04	16.56	20.95	21.79	23.71	19.86	13.72	7.90	4.43	2.68
$\bar{K}_T$	.31	.35	.47	.52	.54	.53	.59	.57	.51	.43	.37	.29
$\bar{T}_a$ C	5.0	7.0	7.0	10.0	13.0	16.0	18.0	18.0	16.0	12.0	8.0	6.0
DD C-DAY	410.	333.	321.	220.	134.	65.	28.	26.	72.	183.	302.	365.
SHREVEPORT LA (LAT. 32.2)												
$\bar{H}$ MJ/m^2	9.45	11.67	15.81	19.53	22.92	22.75	23.50	21.75	17.36	14.47	10.16	8.28
$\bar{K}_T$	.48	.48	.52	.54	.58	.56	.58	.58	.54	.55	.49	.45
$\bar{T}_a$ C	8.0	10.0	14.0	18.0	22.0	26.0	28.0	28.0	25.0	19.0	13.0	9.0
DD C-DAY	307.	237.	169.	45.	0.	0.	0.	0.	0.	26.	165.	265.
SILVER HILL MD (LAT. 38.5)												
$\bar{H}$ MJ/m^2	7.61	10.20	14.22	18.32	21.45	23.21	21.58	19.19	16.60	12.34	8.45	6.82
$\bar{K}_T$	.48	.48	.51	.53	.54	.56	.54	.53	.55	.53	.49	.47
$\bar{T}_a$ C	2.0	3.0	7.0	13.0	18.0	23.0	25.0	24.0	21.0	15.0	9.0	3.0
DD C-DAY	506.	431.	343.	147.	40.	0.	0.	0.	8.	106.	283.	476.

	JAN	FEB	MAR	APR	MAY	JUNE	JULY	AUG	SEP	OCT	NOV	DEC
SPOKANE	WA (LAT. 47.4)											
$\bar{H}$ MJ/m^2	4.94	8.91	13.26	19.53	23.25	25.30	27.73	23.17	17.02	8.62	5.48	3.18
$\bar{K}_T$	.47	.56	.56	.61	.60	.61	.70	.67	.64	.47	.46	.35
$\bar{T}_a$ C	-3.0	.0	3.0	8.0	13.0	16.0	21.0	20.0	15.0	9.0	2.0	-1.0
DD C-DAY	682.	510.	474.	315.	182.	80.	12.	26.	109.	296.	492.	620.
SPRINGFIELD	MO (LAT. 37.1)											
$\bar{H}$ MJ/m^2	8.33	11.39	15.41	18.59	22.11	23.91	23.61	22.15	19.01	13.94	99.22	7.54
$\bar{K}_T$	.50	.52	.54	.53	.56	.58	.59	.60	.62	.59	5.52	.49
$\bar{T}_a$ C	1.0	3.0	7.0	14.0	18.0	23.0	25.0	25.0	21.0	15.0	7.0	2.0
DD C-DAY	553.	436.	367.	153.	52.	6.	0.	3.	19.	126.	325.	499.
STATE COLLEGE	PA (LAT. 40.5)											
$\bar{H}$ MJ/m^2	5.81	8.45	12.42	15.60	19.53	22.75	22.08	18.99	15.10	11.50	6.48	5.02
$\bar{K}_T$	.3[illegible]	.42	.46	.46	.50	.55	.55	.53	.51	.52	.41	.38
$\bar{T}_a$ C	-2.0	-2.0	2.0	9.0	15.0	20.0	22.0	21.0	17.0	11.0	5.0	-1.0
DD C-DAY	654.	572.	489.	267.	115.	13.	3.	8.	61.	214.	400.	609.
STILLWATER	OK (LAT. 36.1)											
$\bar{H}$ MJ/m^2	8.66	11.96	16.23	19.07	20.99	24.80	24.80	22.67	19.03	14.72	10.75	8.53
$\bar{K}_T$	.50	.53	.56	.54	.53	.60	.62	.62	.61	.60	.58	.53
$\bar{T}_a$ C	2.0	5.0	9.0	16.0	20.0	25.0	27.0	27.0	22.0	17.0	9.0	4.0
DD C-DAY	481.	358.	287.	97.	21.	0.	0.	0.	6.	81.	258.	429.
SUMMIT	MT (LAT. 48.2)											
$\bar{H}$ MJ/m^2	5.11	6.78	11.22	17.33	19.34	20.64	23.45	21.35	14.82	9.04	4.27	3.18
$\bar{K}_T$	.51	.44	.48	.55	.50	.50	.59	.62	.56	.51	.38	.37
$\bar{T}_a$ C	-8.0	-5.0	-4.0	1.0	6.0	10.0	14.0	13.0	8.0	4.0	-2.0	-6.0
DD C-DAY	854.	691.	715.	518.	377.	253.	143.	171.	302.	457.	647.	777.
SYRACUSE	NY (LAT. 43.1)											
$\bar{H}$ MJ/m^2	5.40	8.08	12.14	15.66	20.26	23.11	23.36	19.72	14.99	10.09	5.19	4.31
$\bar{K}_T$	.41	.43	.47	.47	.52	.56	.58	.55	.53	.49	.36	.37
$\bar{T}_a$ C	-5.0	-4.0	1.0	8.0	14.0	19.0	22.0	21.0	17.0	11.0	5.0	-2.0
DD C-DAY	713.	628.	548.	308.	151.	26.	6.	10.	67.	218.	400.	636.

	JAN	FEB	MAR	APR	MAY	JUNE	JULY	AUG	SEP	OCT	NOV	DEC
TALLAHASSEE FL (LAT. 30.3)												
$\overline{H}$ MJ/m^2	10.33	13.01	17.69	20.20	22.92	19.91	22.75	22.46	17.73	14.76	15.22	13.01
$\overline{K}_T$	.49	.51	.57	.55	.58	.49	.57	.60	.54	.54	.69	.67
$\overline{T}_a$ C	11.0	12.0	16.0	19.0	23.0	26.0	27.0	27.0	25.0	20.0	15.0	12.0
DD C-DAY	227.	179.	104.	19.	0.	0.	0.	0.	0.	17.	113.	209.
TAMPA FL (LAT. 27.6)												
$\overline{H}$ MJ/m^2	13.67	16.35	19.95	22.79	24.88	23.96	22.29	20.70	18.99	16.94	14.93	12.63
$\overline{K}_T$	.61	.61	.62	.62	.63	.59	.56	.55	.56	.60	.64	.60
$\overline{T}_a$ C	16.0	16.0	19.0	21.0	24.0	26.0	27.0	27.0	26.0	23.0	19.0	16.0
DD C-DAY	113.	98.	50.	5.	0.	0.	0.	0.	0.	0.	39.	94.
TRENTON NJ (LAT. 40.1)												
$\overline{H}$ MJ/m^2	7.24	10.22	14.36	17.75	20.56	22.86	22.61	19.64	16.29	12.31	8.16	6.49
$\overline{K}_T$	.48	.50	.53	.52	.52	.55	.56	.54	.55	.55	.50	.48
$\overline{T}_a$ C	.0	.0	5.0	11.0	16.0	21.0	24.0	23.0	19.0	13.0	7.0	1.0
DD C-DAY	567.	492.	410.	213.	75.	0.	0.	0.	22.	140.	312.	518.
TUCSON AZ (LAT. 32.1)												
$\overline{H}$ MJ/m^2	13.09	16.81	22.83	27.85	30.95	29.65	26.26	24.67	24.30	18.69	14.85	12.42
$\overline{K}_T$	.66	.68	.75	.77	.78	.72	.65	.66	.75	.71	.71	.67
$\overline{T}_a$ C	10.0	11.0	14.0	18.0	22.0	27.0	30.0	28.0	26.0	20.0	14.0	10.0
DD C-DAY	262.	191.	134.	42.	3.	0.	0.	0.	0.	14.	128.	226.
TULSA OK (LAT. 36.1)												
$\overline{H}$ MJ/m^2	8.67	11.43	15.45	18.30	21.56	24.41	23.86	22.23	18.34	13.77	9.80	8.16
$\overline{K}_T$	.50	.51	.54	.52	.55	.59	.59	.60	.59	.57	.53	.51
$\overline{T}_a$ C	3.0	5.0	9.0	16.0	20.0	25.0	28.0	27.0	23.0	17.0	10.0	4.0
DD C-DAY	489.	370.	293.	98.	16.	0.	0.	0.	6.	79.	260.	434.
TWIN FALLS ID (LAT. 40.3)												
$\overline{H}$ MJ/m^2	6.82	10.05	14.86	19.34	23.11	24.79	25.20	22.61	18.09	11.97	7.37	5.49
$\overline{K}_T$	.46	.50	.55	.57	.59	.60	.63	.63	.61	.54	.46	.41
$\overline{T}_a$ C	-1.0	1.0	4.0	9.0	13.0	17.0	22.0	21.0	16.0	10.0	4.0	.0
DD C-DAY	644.	490.	454.	290.	161.	73.	0.	12.	99.	260.	442.	589.

		JAN	FEB	MAR	APR	MAY	JUNE	JULY	AUG	SEP	OCT	NOV	DEC
WASHINGTON		DC (LAT. 38.5)											
$\overline{H}$	MJ/m^2	6.65	9.62	13.38	16.85	18.69	23.34	22.12	19.32	15.35	11.75	8.82	6.15
$\overline{K}_T$		.42	.45	.48	.49	.47	.56	.55	.53	.51	.51	.51	.42
$\overline{T}_a$	C	2.0	3.0	7.0	13.0	18.0	23.0	25.0	24.0	21.0	15.0	9.0	3.0
DD	C-DAY	484.	423.	348.	160.	41.	0.	0.	0.	18.	121.	288.	463.
WICHITA		KS (LAT. 37.4)											
$\overline{H}$	MJ/m^2	[illegible].29	11.97	15.99	19.76	22.78	25.20	24.41	22.57	18.71	14.40	10.26	8.29
$\overline{K}_T$		.56	.55	.56	.57	.58	.61	.61	.62	.61	.61	.58	.54
$\overline{T}_a$	C	.0	2.0	6.0	14.0	19.0	24.0	27.0	26.0	21.0	15.0	7.0	1.0
DD	C-DAY	581.	447.	373.	153.	50.	4.	0.	0.	18.	117.	337.	526.
YUMA		AZ (LAT. 32.4)											
$\overline{H}$	MJ/m^2	12.77	16.79	21.64	26.50	29.43	29.52	27.30	24.58	22.19	18.50	13.82	11.35
$\overline{K}_T$		.65	.69	.71	.74	.74	.72	.68	.66	.69	.71	.67	.62
$\overline{T}_a$	C	12.0	15.0	17.0	21.0	25.0	29.0	34.0	33.0	30.0	24.0	17.0	13.0
DD	C-DAY	171.	107.	54.	13.	0.	0.	0.	0.	0.	0.	60.	153.
AKLAVIK		NW (LAT. 68.1)											
$\overline{H}$	MJ/m^2	.21	2.09	8.45	16.23	21.54	22.08	18.65	12.42	6.52	2.59	.46	.04
$\overline{K}_T$		1.61	.56	.72	.69	.62	.54	.49	.44	.41	.44	.65	.00
$\overline{T}_a$	C	-28.0	-27.0	-22.0	-12.0	.0	9.0	13.0	10.0	3.0	-7.0	-19.0	-26.0
DD	C-DAY	1462.	1298.	1268.	930.	591.	268.	152.	255.	448.	786.	1147.	1406.
CHURCHILL		MA (LAT. 58.4)											
$\overline{H}$	MJ/m^2	2.72	6.27	12.75	18.61	21.33	22.16	21.12	15.89	9.41	4.81	2.51	1.46
$\overline{K}_T$		.65	.67	.73	.67	.58	.54	.55	.51	.44	.41	.47	.49
$\overline{T}_a$	C	-27.0	-26.0	-19.0	-10.0	-2.0	5.0	11.0	11.0	5.0	-1.0	-11.0	-21.0
DD	C-DAY	1421.	1265.	1183.	872.	641.	375.	200.	208.	378.	601.	900.	1249.
EDMONTON		AT (LAT. 53.3)											
$\overline{H}$	MJ/m^2	3.72	7.36	13.05	17.27	21.29	21.45	22.00	17.10	12.46	7.86	4.64	2.76
$\overline{K}_T$		.53	.59	.64	.58	.57	.52	.56	.52	.52	.53	.56	.48
$\overline{T}_a$	C	-14.0	-11.0	-5.0	4.0	11.0	14.0	16.0	15.0	10.0	5.0	-4.0	-10.0
DD	C-DAY	1006.	844.	739.	425.	222.	123.	41.	100.	228.	410.	675.	891.

	JAN	FEB	MAR	APR	MAY	JUNE	JULY	AUG	SEP	OCT	NOV	DEC
KAPUSKASING OT (LAT. 49.2)												
$\overline{H}$ MJ/m^2	4.60	7.95	12.96	15.47	17.15	20.07	20.07	16.73	11.29	6.69	3.35	3.35
$\overline{K}_T$	.49	.53	.57	.49	.45	.49	.51	.49	.44	.39	.31	.42
$\overline{T}_a$ C	-18.0	-16.0	-9.0	.0	7.0	14.0	16.0	15.0	10.0	4.0	-4.0	-14.0
DD C-DAY	1132.	964.	868.	543.	322.	123.	41.	95.	225.	420.	692.	1004.
LETHBRIDGE AT (LAT. 49.4)												
$\overline{H}$ MJ/m^2	5.02	8.78	14.22	17.56	21.75	24.25	25.51	21.75	15.47	10.04	5.86	3.76
$\overline{K}_T$	.54	.59	.63	.56	.57	.59	.64	.64	.60	.59	.55	.47
$\overline{T}_a$ C	-8.0	-7.0	-2.0	5.0	11.0	14.0	17.0	16.0	12.0	7.0	.0	-4.0
DD C-DAY	832.	717.	644.	387.	224.	118.	31.	62.	177.	339.	562.	709.
MONCTON NB (LAT. 46.1)												
$\overline{H}$ MJ/m^2	4.18	7.53	12.13	15.89	18.40	18.82	19.65	17.15	12.96	8.78	4.60	3.76
$\overline{K}_T$	.37	.45	.50	.49	.48	.46	.49	.49	.48	.46	.36	.38
$\overline{T}_a$ C	-8.0	-8.0	-3.0	3.0	9.0	15.0	17.0	16.0	13.0	7.0	1.0	-5.0
DD C-DAY	823.	742.	663.	438.	260.	95.	34.	58.	153.	339.	495.	746.
MONTREAL QU (LAT. 45.3)												
$\overline{H}$ MJ/m^2	4.60	8.36	13.38	16.73	19.65	20.49	21.33	18.40	12.96	8.36	4.18	3.35
$\overline{K}_T$	.39	.48	.54	.51	.51	.50	.53	.52	.47	.43	.32	.32
$\overline{T}_a$ C	-9.0	-7.0	-2.0	5.0	12.0	17.0	18.0	17.0	15.0	8.0	1.0	-6.0
DD C-DAY	870.	767.	653.	380.	176.	38.	5.	24.	92.	289.	490.	773.
OTTAWA OT (LAT. 45.3)												
$\overline{H}$ MJ/m^2	6.02	9.53	14.01	16.85	20.78	23.34	22.87	19.61	14.85	9.24	5.14	4.56
$\overline{K}_T$	.51	.55	.57	.51	.54	.56	.57	.56	.54	.48	.39	.44
$\overline{T}_a$ C	-10.0	-10.0	-3.0	5.0	12.0	16.0	17.0	16.0	14.0	8.0	.0	-8.0
DD C-DAY	902.	801.	684.	393.	189.	50.	14.	45.	123.	315.	520.	816.
ST.JOHNS NF (LAT. 47.3)												
$\overline{H}$ MJ/m^2	3.35	6.27	10.04	13.38	16.73	17.98	18.40	14.22	11.71	7.11	3.35	2.93
$\overline{K}_T$	.32	.39	.42	.42	.43	.44	.46	.41	.44	.39	.28	.32
$\overline{T}_a$ C	-4.0	-4.0	-2.0	1.0	5.0	10.0	15.0	15.0	11.0	6.0	2.0	-1.0
DD C-DAY	701.	650.	659.	515.	394.	240.	103.	100.	190.	362.	462.	618.

	JAN	FEB	MAR	APR	MAY	JUNE	JULY	AUG	SEP	OCT	NOV	DEC
TORONTO OT (LAT. 43.4)												
$\bar{H}$ MJ/m^2	5.06	7.74	12.21	15.56	19.99	21.75	22.04	18.15	14.39	9.37	5.14	3.97
$\bar{K}_T$	.39	.42	.48	.47	.51	.53	.55	.51	.51	.46	.36	.34
$\bar{T}_a$ C	-3.0	-3.0	.0	6.0	12.0	17.0	18.0	17.0	15.0	10.0	4.0	-1.0
DD C-DAY	685.	622.	563.	342.	166.	34.	4.	10.	84.	244.	422.	617.
VANCOUVER BC (LAT. 48.6)												
$\bar{H}$ MJ/m^2	3.18	4.35	7.82	14.34	19.57	20.16	22.79	16.64	10.54	6.69	3.97	2.34
$\bar{K}_T$	.32	.28	.34	.45	.51	.49	.57	.49	.40	.38	.36	.28
$\bar{T}_a$ C	2.0	3.0	6.0	9.0	12.0	15.0	16.0	16.0	14.0	10.0	6.0	4.0
DD C-DAY	479.	402.	376.	278.	172.	87.	45.	48.	122.	253.	365.	437.
WINNIPEG MA (LAT. 49.5)												
$\bar{H}$ MJ/m^2	5.48	9.41	15.18	18.36	21.33	21.95	23.75	19.74	13.34	8.57	4.98	3.85
$\bar{K}_T$	.59	.64	.67	.59	.56	.53	.60	.58	.52	.50	.47	.49
$\bar{T}_a$ C	-17.0	-15.0	-7.0	3.0	11.0	15.0	17.0	17.0	12.0	6.0	-4.0	-13.0
DD C-DAY	1116.	955.	814.	452.	225.	82.	21.	39.	179.	379.	695.	976.

APPENDIX 3
UNIT CONVERSION TABLES

SI UNITS

Basic Units

meter	m	length
kilogram	kg	mass
second	s	time
Kelvin	K	temperature

Derived Units

liter	l	volume	$10^{-3}m^3$
Newton	N	force	kg-m/s
Joule	J	energy	N-m
Watt	W	power	J/s
Hour	hr	time	3600 s

Decimal Multiples of Units

tera	T	10^{12}
giga	G	10^{9}
mega	M	10^{6}
kilo	k	10^{3}
milli	m	10^{-3}
micro	u	10^{-6}
nano	n	10^{-9}
pico	p	10^{-12}

UNIT CONVERSIONS

Length

1 ft = 0.3048 m
1 mile = 1.6093 m
1 inch = 25.4 mm
1 yard = 0.9144 m

Velocity

1 ft/min = 0.00508 m/s
1 mile/hr = 0.44704 m/s

Area

1 ft^2 = 0.092903 m^2
1 $mile^2$ = 2.58999 km^2
1 $inch^2$ = 0.000645 m2

Volume	
1 ft^3	= 28.3168 l
1 gal	= 3.78544 l
1 ft^3	= 7.48 gal
1 $yard^3$	= 0.7645 m^3
1 gal/ft^2	= 0.02454 l/m^2

Volumetric Rate	
1 cfm	= 0.47195 l/s
1 gal/min	= 0.06309 l/s
1 $gal/min\text{-}ft^2$	= 0.6791 $l/s\text{-}m^2$
1 cfm/ft^2 (air)	= 0.1968 $l/s\text{-}m^2$

Mass	
1 lb	= 0.453492 kg
1 oz	= 28.3495 g

Mass Flow Rate	
1 lb/hr	= 0.000126 kg/s
1 $lb/hr\text{-}ft^2$	= 0.001356 $kg/s\text{-}m^2$

Temperature

Scales

F = C x 1.8 + 32
C = (F-32) x 5/9
K = C + 273
R = F + 462

Differences

1 F = 0.55556 C
1 C = 1.8 F

Energy	
1 BTU	= 1.05506 kJ
1 Therm	= 105.506 MJ
1 cal	= 4.1868 J
1 kW-hr	= 3.6 MJ
1 langley	= 41.86 kJ/m^2

Power	
1 BTU/hr	= 0.29307 W
1 ton (refg)	= 3.51685 kW
1 kcal/hr	= 1.163 W
1 hp	= 0.74570 kW

Energy Flux	
1 $BTU/hr\text{-}ft^2$	= 3.15469 W/m^2
1 langley/hr	= 11.6277 W/m^2
1 $cal/cm^2\text{-}min$	= 697.4 W/m^2
1 $BTU/hr\text{-}ft^2\text{-}F$	= 5.67826 $W/m^2\text{-}C$
1 BTU/hr-ft-F	= 1.70307 W/m-C

NUMERICAL VALUES OF SOME PROPERTIES

Solar Constant = 1353 W/m^2
= 1.940 langleys/min
= 428 $BTU/hr\text{-}ft^3$

	Density	Specific Heat
Air	1.204 kg/m	1012 J/kg-C
	0.07516 lb/ft^3	0.241 BTU/lb-F
Water	1000 kg/m^3	4190 J/kg-C
	62.42 lb/ft^3	1.00 BTU/lb-F
	8.34 lb/gal	
Rock	2400 kg/m^3	838 J/kg-C
	150 lb/ft^3	0.2 BTU/lb-F
Antifreeze (50-50 ethylene glycol/water)	1065 kg/m^3	3350 J/kg-C
	66.50 lb/ft^3	0.80 BTU/lb-F

APPENDIX 4
GLOSSARY

ABSORBER PLATE - the surface in a flat-plate collector upon which incident solar radiation is absorbed.

ABSORPTANCE - the ratio of the radiation absorbed by a surface to that incident on the surface.

ACTIVE SOLAR HEATING SYSTEM - a solar heating system which uses specialized equipment to collect, store and distribute solar heat in a controlled manner.

AIR HEATING SYSTEM - a solar heating system in which air is heated in the solar collector and used as the energy transfer medium to the rest of the system.

ANNUAL LOAD FRACTION - fraction of the annual heating needs supplied by solar energy.

AUXILIARY ENERGY - energy supplied for heating by some means other than solar energy (oil, gas, or electricity).

BEAM RADIATION - solar radiation which is not scattered by dust or water droplets. It is capable of being focussed and casts shadows.

BLACK - a property of a substance which has a high absorptance for radiation.

BUILDING OVERALL ENERGY LOSS COEFFICIENT-AREA PRODUCT - the factor which when multiplied by the monthly degree-days yields the monthly space heating load.

CAPACITANCE RATE - mass flow rate times specific heat of the fluid flowing through a component such as a heat exchanger.

COLLECTOR EFFICIENCY - the ratio of the useful energy gain for a time period to the solar energy incident on the surface during the same time period.

COLLECTOR-HEAT EXCHANGER CORRECTION FACTOR, F_R'/F_R - an index ranging in value from 0 to 1 indicating the penalty in useful energy collection resulting from using heat exchange between the collector and the storage tank in liquid solar heating systems.

COLLECTOR HEAT REMOVAL EFFICIENCY FACTOR, F_R - the ratio of the actual useful energy gain of a flat-plate solar collector to the energy gain if the entire collector plate were at the temperature of the inlet fluid.

COLLECTOR OVERALL ENERGY LOSS COEFFICIENT, U_L - a parameter characterizing the energy losses of the collector to the surroundings.

COMPOUND PARABOLIC CONCENTRATORS - a type of concentrating collector using parabolic reflectors which does not form an image of the sun on the receiving surface.

CONCENTRATING SOLAR COLLECTOR - a device which focuses beam radiation so as to obtain energy at higher temperatures than attainable with flat-plate solar collectors.

CONTROLS - thermostats and temperature sensing devices used to manipulate fans, pumps, and dampers in solar heating systems.

DECLINATION - the angular position of the sun at solar noon with respect to the plane of the equator, i.e., the angular position of the sun north or south of the equator; a function of the time of year.

DEGREE-DAYS (monthly) - the sum of the differences between 18.3 C (65 F) and the mean daily temperature for each day of the month.

DESIGN HEATING LOAD - the maximum probable space heating needs of a building.

DESIGN TEMPERATURE DIFFERENCE - the maximum probable difference between the indoor and the ambient temperatures.

DIFFUSE RADIATION - solar radiation which is scattered by air molecules, dust or water droplets before reaching the ground and not capable of being focused.

DIMENSIONLESS VARIABLE - a quantity which does not have dimensional units and is therefore has the same value in any system of units.

DISCOUNTED CASH FLOW - present worth of a future payment.

DOMESTIC HOT WATER - hot water used for conventional purposes such as bathing and washing.

DOMESTIC WATER HEATING SYSTEM - a solar heating system which supplies a portion of the energy needed to heat water used for domestic purposes, such as bathing and washing.

EFFECTIVENESS - the ratio of actual heat transfer in a heat exchanger to the maximum possible heat transfer.

EMITTANCE - the ratio of the radiant energy (heat) emitted from a surface at a given temperature to the energy emitted by a perfect black body at the same temperature.

EXTRATERRESTRIAL RADIATION - the solar radiation which would be received on a horizontal surface if there were no atmosphere.

f-CHART - a correlation, presented graphically and analytically, which expresses the monthly load fraction supplied by solar energy in terms of two dimensionless variables which include measured collector parameters and monthly average meteorological conditions.

FCHART - an interactive computer program available from the Solar Energy Laboratory, University of Wisconsin - Madison, which calculates solar heating system performance and economics by the methods described in this book.

FLAT-PLATE SOLAR COLLECTOR - the basic heat collection device used in solar heating systems; consists of a "black" plate, insulated on the bottom and edges, and covered by one or more transparent covers.

FRESNEL COLLECTOR - a concentrating solar collector which focuses solar radiation using a Fresnel lens.

HOUR ANGLE - 15° times the number of hours from solar noon.

INCIDENCE ANGLE - the angle between the perpendicular to a surface and the direction of the solar radiation.

LIFE-CYCLE ECONOMIC ANALYSIS - a method of determining all future costs in terms of today's dollars.

LIQUID-BASED SOLAR HEATING SYSTEM - a solar heating system in which liquid, either water or an antifreeze solution, is heated in the solar collectors.

LIQUID-TO-LIQUID HEAT EXCHANGER - a device for heat exchange between two liquid streams.

LOAD - space or domestic water heating needs which is to be supplied by solar or conventional energy.

LONG-WAVE RADIATION - infrared or radiant heat.

MARKET DISCOUNT RATE - the rate of return which can be expected on the best investment.

MEAN DAILY TEMPERATURE - average of the minimum and maximum daily temperatures used to determine the number of degree-days.

MIN[a,b] - smaller of the values enclosed in brackets.

NORMAL RADIATION - the component of solar radiation which is perpendicular to the absorbing surface.

PARABOLIC FOCUSING COLLECTOR - a concentrating collector which focuses beam radiation with a parabolic reflector.

PASSIVE SOLAR HEATING - solar heating of a building accomplished by architectural design without the aid of mechanical equipment.

PEBBLE BED - a large bin of uniform size pebbles used for storing solar heat in solar air heating systems.

PRESENT WORTH - the amount of money which must be invested today in order to have a specified amount at a future time.

PYRANOMETER - an instrument for measuring total (beam, diffuse, and reflected) solar radiation.

REFLECTED RADIATION - solar radiation which is reflected from a surface such as the ground (and, as used here, is ultimately incident on the collector surface).

REFLECTANCE - the ratio of radiation reflected from a surface to the total radiation incident on the surface.

RETURN ON INVESTMENT - the market discount rate which will result in zero life cycle savings.

SELECTIVE SURFACE - a surface which has a high absorptance for solar radiation, but a low emittance for thermal (long-wave) radiation.

SOLAR SAVINGS - the life cycle cost of conventional heating minus the life cycle cost of solar heating.

STORAGE CAPACITY - the amount of energy which can be stored by a solar heating system to be used at a later time for space or water heating.

SUNSET HOUR ANGLE - the hour angle occurring at sunset (See HOUR ANGLE)

TRANSMITTANCE - the ratio of the radiation passing through a material to the radiation incident on the upper surface of that material.

TRANSMITTANCE-ABSORPTANCE PRODUCT - the product of the transmittance of the transparent collector cover(s) and the absorptance of the collector plate.

UNIFORM ANNUAL PAYMENT - the payment to be made each year on a mortgage loan.

USEFUL ENERGY GAIN - the energy collected by a solar collector which is not lost to the surroundings and can ultimately be used for space or water heating.

WATER-AIR HEAT EXCHANGER - a device in which air is either heated or cooled by flowing water.

APPENDIX 5
BLANK WORKSHEETS

COLLECTOR ORIENTATION WORKSHEET 1

AVERAGE DAILY RADIATION ON TILTED SURFACES

A. Location ____________ B. Latitude ϕ = ____________ C. Inclination s = ____________

D. (1+cos s)/2 ____________ E. Ground Reflectance ρ = ____________ F. ρ(1-cos s)/2 = ____________

G1. Month	G2. $\overline{H}$ J/Day-m^2 (Appendix 2)	G3. $\overline{K}_T$ (Appendix 2)	G4. $\overline{H}_d/\overline{H}$ Fig. 3.1 or Eqn. 3.3)	G5. $1-\overline{H}_d/\overline{H}$ (1-G4.)	G6. $\overline{R}_b$ (Fig. 3.2 or Eqn. 3.4)	G7. Beam (G5.xG6.)	G8. Diffuse (D.xG4.)	G9. $\overline{R}$ (G7.+G8.+F.)	G10. $\overline{H}_T$ J/Day-m^2 (G9.xG2.)
Jan	$\times 10^6$								$\times 10^6$
Feb	$\times 10^6$								$\times 10^6$
Mar	$\times 10^6$								$\times 10^6$
Apr	$\times 10^6$								$\times 10^6$
May	$\times 10^6$								$\times 10^6$
Jun	$\times 10^6$								$\times 10^6$
Jul	$\times 10^6$								$\times 10^6$
Aug	$\times 10^6$								$\times 10^6$
Sep	$\times 10^6$								$\times 10^6$
Oct	$\times 10^6$								$\times 10^6$
Nov	$\times 10^6$								$\times 10^6$
Dec	$\times 10^6$								$\times 10^6$

COLLECTOR ORIENTATION WORKSHEET 2

MONTHLY AVERAGE TRANSMITTANCE - ABSORPTANCE PRODUCT

H1. Month	H2. $\overline{\theta}_b$ (Fig.III.5)	H3. τ/τ_n @ θ_b (Fig.III.3)	H4. α/α_n @ θ_b (Fig.III.4)	H5. $\overline{R}_b/\overline{R}$ (G9./G6.)	H6. Beam (G5.xH3.x H4.xH5.)	H7. τ/τ_n @ 60° (Fig.III.3)	H8. Diffuse (D.xG4./G9. x0.92xH7.)	H9. Reflected (F./G9.x0.92 xH7.)	H10. $(\overline{\tau\alpha})/(\tau\alpha)_n$ (H7.+H8.+H9.)
Jan									
Feb									
Mar									
Apr									
May									
Jun									
Jul									
Aug									
Sep									
Oct									
Nov									
Dec									

F-CHART WORKSHEET 1

HEATING LOADS

A. $UA = \frac{\text{Design Space Heating Load [W]}}{\text{Design Temperature Difference [°C]}}$ = ________ [W/°C] (See Section 4.2)

B. Water Usage = ________ [liters/day] x 4190 x $(T_w - T_m)$[J/liter] = ________ [J/Day]

	C1.	C2.	C3.	C4.	C5.
Month	Days Per Mo.	Heating Degree Days [°C-day] (from Appendix 2)	Space Heating Load [J/Month] (86400)(A.)(C2.)	Domestic Water Load [J/Month] (See Section 4.3) (B.)(C1.)	Total Load [J/Month] (C3.)+(C4.)
Jan	31				
Feb	28				
Mar	31				
Apr	30				
May	31				
Jun	30				
Jul	31				
Aug	31				
Sep	30				
Oct	31				
Nov	30				
Dec	31				
Totals	365				

F-CHART WORKSHEET 2

ITEMS MAKING UP X AND Y

C. $F_R U_L (F'_R/F_R)$ = ______________ $[W/m^2 °C]$ (See Sections 2.3 and 2.4)

D. $F_R (\tau\alpha)_n (F'_R/F_R)$= ______________ (See Sections 2.3 and 2.4)

	C6.	C7.	C8.	C9.	C10.	C11.
	Seconds Per Month	$(100-\overline{T}_a)$[°C] ($\overline{T}_a$ can be found in Appendix 2	X/A$[1/m^2]$ $\frac{(C.)(C7.)(C6.)}{(C5.)}$	$(\overline{\tau\alpha})/(\tau\alpha)_n$ (See Sections 3.4 and 3.5)	Daily Radiation on Collector $[J/m^2\text{-Day}]$ (See Section 3.2)	Y/A$[1/m^2]$ $\frac{(D.)(C9.)(C10.)(C1.)}{(C5.)}$
Jan	2.68×10^6				$\times 10^6$	
Feb	2.42×10^6				$\times 10^6$	
Mar	2.68×10^6				$\times 10^6$	
Apr	2.59×10^6				$\times 10^6$	
May	2.68×10^6				$\times 10^6$	
Jun	2.59×10^6				$\times 10^6$	
Jul	2.68×10^6				$\times 10^6$	
Aug	2.68×10^6				$\times 10^6$	
Sep	2.59×10^6				$\times 10^6$	
Oct	2.68×10^6				$\times 10^6$	
Nov	2.59×10^6				$\times 10^6$	
Dec	2.68×10^6				$\times 10^6$	

F-CHART WORKSHEET 3

SOLAR HEATING LOAD FRACTION

E. Storage size correction factor (X/X_o) = ______________

F. Load heat exchanger correction factor (Y/Y_o) = ______________ (=1.0 for air systems)

G. Collector air flow rate correction factor (X/X_o) = ______________ (=1.0 for liquid systems)

	C12. Corrected X/A (C8.)(E.)(G.)	C13. Corrected Y/A (C11.)(F.)	Area = ______ C14. X	C15. Y	C16. f	C17. (C16.)(C5.)	Area = ______ C14. X	C15. Y	C16. f	C17. (C16.)(C5.)	Area = ______ C14. X	C15. Y	C16. f	C17. (C16.)C5.)
Jan														
Feb														
Mar														
Apr														
May														
Jun														
Jul														
Aug														
Sep														
Oct														
Nov														
Dec														
					Totals									

Annual Fractions by Solar
(Total, C17.)/(Total, C5.) = ______________ = ______________ = ______________

F-CHART WORKSHEET 4
ECONOMIC PARAMETERS

H.	Annual mortgage interest rate	______%/100
I.	Term of mortgage	______Yrs.
J.	Down payment (as fraction of investment)	______%/100
K.	Collector area dependent costs	______$\$/m^2$
L.	Area independent costs	______$\$$
M.	Present cost of solar backup system fuel	______$\$$/GJ
N.	Present cost of conventional system fuel	______$\$$/GJ
O.	Efficiency of solar backup furnace	______%/100
P.	Efficiency of conventional system furnace	______%/100
Q.	Property tax rate (as fraction of investment)	______%/100
R.	Effective income tax bracket (state+federal-state x federal)	______%/100
S.	Extra ins. & maint. costs (as fraction of investment)	______%/100
T.	General inflation rate per year	______%/100
**U.	Fuel inflation rate per year	______%/100
V.	Discount rate (after tax return on best alternative investment)	______%/100
W.	Term of economic analysis	______Yrs.
X.	First year non-solar fuel expense (total, C5)(N.)/(P.)÷10^9	______$\$$
*Y.	Depreciation lifetime	______Yrs.
Z.	Salvage value (as fraction of investment)	______%/100
AA.	Table 6.2 with Yr = (W.), Column = (U.) and Row = (V.)	______
BB.	" (W.) " (T.) " (V.)	______
CC.	" MIN(I.,W.) " (H.) " (V.)	______
†*DD.	" MIN(W.,Y.) " (Zero) " (V.)	______
EE.	" (I.) " (Zero) " (H.)	______
FF.	" MIN(I.,W.) " (Zero) " (V.)	______
GG.	(FF.)/(EE.), Loan payment	______
HH.	(GG.)+(CC.)[(H.)-1/(EE.)], Loan interest	______
II.	(J.)+(1-J.)[(GG.)-(HH.)(R.)], Capital cost	______
JJ.	(S.)(BB.), I&M cost	______
KK.	(Q.)(BB)(1-R.), Property tax	______
LL.	$(Z.)/(1+V.)^{(W.)}$, Salvage value	______
*MM.	(R.)(DD.)(1-Z.)/(Y.), Depreciation	______
NN.	Other costs (see Section 6.9)	______
OO.	(II.)+(JJ.)+(KK.)-(LL.)+(NN.), Residential costs	______
PP.	(II.)+(JJ.)(1-R.)+(KK.)-(LL.)-(MM.)+(NN.)(1-R.), Commercial costs	______

**For other fuel inflation factors see Section 6.9.
*Commercial only.
†Straight line only. Use Tables 6.3A or 6.3B for other depreciation methods.

F-CHART WORKSHEET 5

ECONOMIC ANALYSIS

R1. Collector Area (Worksheet 3)				
R2. Fraction by Solar (Worksheet 3)				
R3. Investment in Solar (K.)(R1.)+(L.)				
R4. 1st Year Fuel Expense (Total, C5)(1-R2.)(M.)/(O.) $\div 10^9$				
R5. Fuel Savings (X.-R4.)(AA.)				
R6. Expenses (Residential) (OO.)(R3.)				
R7. Expenses (Commercial) (PP.)(R3.)				
R8. Savings (Residential) (R5.)-(R6.)				
R9. Savings (Commercial) (R5.)(1-R.)-(R7.)				

F-CHART WORKSHEET 6

YEARLY SAVINGS FOR COLLECTOR AREA = __________

R10.	Year (n) (first year=1)				
†R11.	Current Mortgage [(R11.)-(R14.)+(R18.)]				
R12.	Fuel Savings $(X.-R4.)(1+U.)^{n-1}$				
R13.	Down Payment (1st year only) (R3.)(J.)				
R14.	Mortgage Payment (1-J.)(R3.)/(EE.)				
R15.	Extra Insurance & Maintenance $(S.)(R3.)(1+T.)^{n-1}$				
R16.	Extra Property Tax $(R3.)(Q.)(1+T.)^{n-1}$				
R17.	Sum (R13.+R14.+R15.+R16.)				
R18.	Interest on Mortgage (R11.)(H.)				
R19.	Tax Savings (R.)(R16.+R18.)				
*R20.	Depreciation (st. line) (R3.)(1-Z.)/(Y.)				
*R21.	Business Tax Savings (R.)(R20.+R15.-R12.)				
R22.	Salvage Value (R2.)(Z.) (Last Year Only)				
R23.	Solar Savings (R12.-R17.+R19.+R21.+R22.)				
**R24.	Discounted Savings $(R23.)/(1+V.)^{n}$				

†For the first year use [(R3.)(1-J.)]; for subsequent years use equation with previous years values.

*Income producing property only.

**The down payment should not be discounted.

APPENDIX 6
THE FCHART INTERACTIVE PROGRAM

An interactive computer program, called FCHART, has been developed to do the calculations described in this text. Included are meteorological data for approximately 170 North American locations. The program is available from the Solar Energy Laboratory, University of Wisconsin, Madison.

The results of the economic calculations are presented in terms of the present worth of the solar energy system and the present worth of the system without solar energy. The present worth of the solar savings, the quantity calculated by the methods of this book, is then the difference between the present worths of the non-solar and solar systems. Sample outputs for a case in which the specified collector area is used and for a case in which the optimized collector area is determined by the program are shown on the following pages.

```
 UNIVERSITY OF WISCONSIN
 SOLAR ENERGY LABORATORY
 FCHART VERSION 2.0

 THIS PROGRAM UTILIZES THE DESIGN CHARTS DEVELOPED AT THIS LABORATORY
 TO SIZE COLLECTORS FOR SOLAR SPACE AND DOMESTIC WATER HEATING SYSTEMS
 OF CONVENTIONAL DESIGN. TO USE, ANSWER THE QUESTIONS.

 DO YOU NEED INSTRUCTIONS(Y,N OR X)?
Y
 "FCHART" IS SO SIMPLE TO USE THAT YOU MAY NEVER ASK FOR INSTRUCTIONS
 AGAIN. RESPOND TO THE ABOVE QUESTION WITH "N" OR "X" TO BYPASS THESE
 INSTRUCTIONS. "N" MEANS NO INSTRUCTIONS ARE NEEDED, BUT LONG VERSIONS
 OF ALL SUBSEQUENT QUESTIONS WILL BE PRINTED OUT. "X" IS FOR THE EXPERI-
 ENCED USER WHO IS IN A HURRY AND WANTS SHORT VERSIONS OF THE QUESTIONS.
      AFTER A SERIES OF QUESTIONS THE PROGRAM WILL ASK YOU TO "TYPE IN
 CODE NUMBER AND NEW VALUE". AT THIS POINT YOU HAVE THE FOLLOWING ALTER-
 NATIVES. TYPE IN:

  "L"   - TO LIST ALL OF THE PARAMETERS DESCRIBING THE SOLAR
  .       SYSTEM ALONG WITH THEIR CODE NUMBERS AND INITIAL VALUES.
  "L N" - TO ONLY LIST PARAMETER NUMBER "N".
  "N V" - TO CHANGE THE VALUE OF PARAMETER "N" TO "V".
  "B"   - TO RETURN TO THE BEGINNING OF THE PROGRAM
  "R"   - TO RUN AN ANALYSIS OF THE SPECIFIED SOLAR SYSTEM.
  "S"   - TO STOP EXECUTION OF THE PROGRAM.

 AFTER TYPING IN "R",THE PARAMETER VALUES ARE SCANNED TO SEE IF THEY ARE
 REASONABLE. IF THE CITY CALL NUMBER IS † THAN 172 YOU HAVE THE OPTION
 OF MODIFYING WEATHER DATA. WHEN OPTIMIZING COLLECTOR AREA,YOU WILL NEED
 TO SPECIFY THE COLLECTOR MODULE SIZE.

 YOU MAY USE EITHER SI OR ENGLISH UNITS.
 DO YOU WISH TO USE SI UNITS?
Y
 WOULD YOU LIKE A LISTING OF LOCATIONS FOR WHICH CALCULATIONS CAN BE
 MADE?
N
 YOU MAY MODEL THE SPACE HEATING LOAD USING THE DEGREE-DAY CONCEPT
 OR YOU MAY TYPE IN A SPACE HEATING LOAD FOR EACH MONTH.
 DO YOU WISH TO USE THE DEGREE-DAY CONCEPT?
Y
 YOU MAY EITHER HAVE THE GROUND REFLECTANCE SET TO 0.2 FOR ALL MONTHS
 OR YOU MAY TYPE IN A VALUE FOR EACH MONTH
 DO YOU WISH TO HAVE THE GROUND REFLECTANCE SET TO 0.2 FOR ALL MONTHS?
Y
 WOULD YOU LIKE THE PROGRAM TO PERFORM AN ECONOMIC ANALYSIS?
Y
 IS THIS AN INCOME PRODUCING BUILDING (NOT A RESIDENCE)?
N
 TYPE IN CODE NUMBER AND NEW VALUE
```

L

CODE	VARIABLE DESCRIPTION	VALUE	UNITS
1	AIR SYSTEM=1,LIQUID SYSTEM=2.................	2.00	
2	COLLECTOR AREA..................................	50.00	M2
3	FRPRIME-TAU-ALPHA PRODUCT(NORMAL INCIDENCE)..	.70	
4	FRPRIME-UL PRODUCT..............................	4.72	W/C-M2
5	NUMBER OF TRANSPARENT COVERS................	2.00	
6	COLLECTOR SLOPE.................................	43.00	DEGREES
7	AZIMUTH ANGLE (E.G. SOUTH=0, WEST=90)........	.00	DEGREES
8	STORAGE CAPACITY................................	315.00	KJ/C-M2
9	EFFECTIVE BUILDING UA..........................	277.78	W/C
10	CONSTANT DAILY BLDG HEAT GENERATION..........	.00	KJ/DAY
11	(EPSILON)(CMIN)/(EFFECTIVE BUILDING UA)......	2.00	
12	HOT WATER USAGE.................................	300.00	L/DAY
13	WATER SET TEMPERATURE..........................	60.00	C
14	WATER MAIN TEMPERATURE........................	11.00	C
15	CITY CALL NUMBER................................	88.00	
16	THERMAL PRINT OUT BY MONTH=1, BY YEAR=2......	1.00	
17	ECONOMIC ANALYSIS ? YES=1, NO=2..............	1.00	
18	USE OPTMZD. COLLECTOR AREA=1, SPECFD. AREA=2.	2.00	
19	SOLAR SYSTEM THERMAL PERFORMANCE DEGRADATION.	.00	%/YR
20	PERIOD OF THE ECONOMIC ANALYSIS..............	20.00	YEARS
21	COLLECTOR AREA DEPENDENT SYSTEM COSTS........	100.00	$/M2 COLL.
22	CONSTANT SOLAR COSTS...........................	1000.00	$
23	DOWN PAYMENT(% OF ORIGINAL INVESTMENT).......	10.00	%
24	ANNUAL INTEREST RATE ON MORTGAGE.............	8.00	%
25	TERM OF MORTGAGE...............................	20.00	YEARS
26	ANNUAL NOMINAL(MARKET) DISCOUNT RATE.........	8.00	%
27	EXTRA INSUR.,MAINT. IN YEAR 1(% OF ORIG.INV.)	1.00	%
28	ANNUAL % INCREASE IN ABOVE EXPENSES..........	6.00	%
29	PRESENT COST OF SOLAR BACKUP FUEL (BF).......	6.00	$/GJ
30	BF RISE: %/YR=1,SEQUENCE OF VALUES=2.........	1.00	
31	IF 1, WHAT IS THE ANNUAL RATE OF BF RISE.....	10.00	%
32	PRESENT COST OF CONVENTIONAL FUEL (CF).......	6.00	$/GJ
33	CF RISE: %/YR=1,SEQUENCE OF VALUES=2.........	1.00	
34	IF 1, WHAT IS THE ANNUAL RATE OF CF RISE.....	10.00	%
35	ECONOMIC PRINT OUT BY YEAR=1, CUMULATIVE=2...	2.00	
36	EFFECTIVE FEDERAL-STATE INCOME TAX RATE......	35.00	%
37	TRUE PROP. TAX RATE PER $ OF ORIGINAL INVEST.	2.00	%
38	ANNUAL % INCREASE IN PROPERTY TAX RATE.......	6.00	%
39	CALC.RT. OF RETURN ON SOLAR INVTMT?YES=1,NO=2	2.00	
40	SALVAGE VALUE (% OF ORIGINAL INVESTMENT).....	.00	%
41	INCOME PRODUCING BUILDING? YES=1,NO=2........	2.00	

TYPE IN CODE NUMBER AND NEW VALUE

R
MADISON WI 43.08

****THERMAL ANALYSIS****

TIME	PERCENT SOLAR	INCIDENT SOLAR (GJ)	HEATING LOAD (GJ)	WATER LOAD (GJ)	DEGREE DAYS (C-DAY)	AMBIENT TEMP (C)
JAN	32.5	19.14	19.92	1.91	830.	-7.
FEB	42.5	20.29	16.69	1.72	696.	-6.
MAR	64.6	27.87	14.39	1.91	599.	0.
APR	82.1	25.61	7.88	1.85	328.	7.
MAY	99.7	27.77	3.96	1.91	165.	13.
JUN	100.0	29.43	.96	1.85	40.	19.
JUL	100.0	31.46	.19	1.91	8.	21.
AUG	100.0	30.01	.52	1.91	22.	20.
SEP	100.0	29.21	2.31	1.85	96.	15.
OCT	90.4	26.10	6.32	1.91	263.	10.
NOV	40.5	16.57	12.12	1.85	505.	1.
DEC	32.5	17.89	17.81	1.91	742.	-5.
YR	56.0	301.36	103.07	22.48	4294.	

****ECONOMIC ANALYSIS****

SPECIFIED COLLECTOR AREA = 50. M2
INITIAL COST OF SOLAR SYSTEM = $ 6000.
THE ANNUAL MORTGAGE PAYMENT FOR 20 YEARS = $ 550.
THE DISCOUNTED PAYBACK PERIOD IS(YR) 12.
YRS UNTIL CUMULATIVE SAVINGS=MORTGAGE PRINCIPLE 15.
PRESENT WORTH OF YEARLY TOTAL COSTS WITH SOLAR = $ 14378.
PRESENT WORTH OF YEARLY TOTAL COSTS W/O SOLAR = $ 16699.
PRESENT WORTH OF CUMULATIVE SOLAR SAVINGS = $ 2322.

TYPE IN CODE NUMBER AND NEW VALUE

18 1
TYPE IN CODE NUMBER AND NEW VALUE
35 1
TYPE IN CODE NUMBER AND NEW VALUE
39 1
TYPE IN CODE NUMBER AND NEW VALUE

L

CODE	VARIABLE DESCRIPTION	VALUE	UNITS
1	AIR SYSTEM=1,LIQUID SYSTEM=2.................	2.00	
2	COLLECTOR AREA................................	50.00	M2
3	FRPRIME-TAU-ALPHA PRODUCT(NORMAL INCIDENCE)..	.70	
4	FRPRIME-UL PRODUCT............................	4.72	W/C-M2
5	NUMBER OF TRANSPARENT COVERS..................	2.00	
6	COLLECTOR SLOPE...............................	43.00	DEGREES
7	AZIMUTH ANGLE (E.G. SOUTH=0, WEST=90)........	.00	DEGREES
8	STORAGE CAPACITY..............................	315.00	KJ/C-M2
9	EFFECTIVE BUILDING UA.........................	277.78	W/C
10	CONSTANT DAILY BLDG HEAT GENERATION..........	.00	KJ/DAY
11	(EPSILON)(CMIN)/(EFFECTIVE BUILDING UA)......	2.00	
12	HOT WATER USAGE...............................	300.00	L/DAY
13	WATER SET TEMPERATURE.........................	60.00	C
14	WATER MAIN TEMPERATURE........................	11.00	C
15	CITY CALL NUMBER..............................	88.00	
16	THERMAL PRINT OUT BY MONTH=1, BY YEAR=2......	1.00	
17	ECONOMIC ANALYSIS ? YES=1, NO=2..............	1.00	
18	USE OPTMZD. COLLECTOR AREA=1, SPECFD. AREA=2.	1.00	
19	SOLAR SYSTEM THERMAL PERFORMANCE DEGRADATION.	.00	%/YR
20	PERIOD OF THE ECONOMIC ANALYSIS..............	20.00	YEARS
21	COLLECTOR AREA DEPENDENT SYSTEM COSTS........	100.00	$/M2 COLL.
22	CONSTANT SOLAR COSTS.........................	1000.00	$
23	DOWN PAYMENT(% OF ORIGINAL INVESTMENT).......	10.00	%
24	ANNUAL INTEREST RATE ON MORTGAGE.............	8.00	%
25	TERM OF MORTGAGE..............................	20.00	YEARS
26	ANNUAL NOMINAL(MARKET) DISCOUNT RATE.........	8.00	%
27	EXTRA INSUR.,MAINT. IN YEAR 1(% OF ORIG.INV.)	1.00	%
28	ANNUAL % INCREASE IN ABOVE EXPENSES..........	6.00	%
29	PRESENT COST OF SOLAR BACKUP FUEL (BF).......	6.00	$/GJ
30	BF RISE: %/YR=1,SEQUENCE OF VALUES=2.........	1.00	
31	IF 1, WHAT IS THE ANNUAL RATE OF BF RISE.....	10.00	%
32	PRESENT COST OF CONVENTIONAL FUEL (CF).......	6.00	$/GJ
33	CF RISE: %/YR=1,SEQUENCE OF VALUES=2.........	1.00	
34	IF 1, WHAT IS THE ANNUAL RATE OF CF RISE.....	10.00	%
35	ECONOMIC PRINT OUT BY YEAR=1, CUMULATIVE=2...	1.00	
36	EFFECTIVE FEDERAL-STATE INCOME TAX RATE......	35.00	%
37	TRUE PROP. TAX RATE PER $ OF ORIGINAL INVEST.	2.00	%
38	ANNUAL % INCREASE IN PROPERTY TAX RATE.......	6.00	%
39	CALC.RT. OF RETURN ON SOLAR INVTMT?YES=1,NO=2	1.00	
40	SALVAGE VALUE (% OF ORIGINAL INVESTMENT).....	.00	%
41	INCOME PRODUCING BUILDING? YES=1,NO=2........	2.00	

TYPE IN CODE NUMBER AND NEW VALUE

R
WHAT IS THE COLLECTOR MODULE SIZE(FT2 OR M2)?
1
MADISON WI 43.08

****THERMAL ANALYSIS****

TIME	PERCENT SOLAR	INCIDENT SOLAR (GJ)	HEATING LOAD (GJ)	WATER LOAD (GJ)	DEGREE DAYS (C-DAY)	AMBIENT TEMP (C)
JAN	26.3	14.93	19.92	1.91	830.	-7.
FEB	34.7	15.83	16.69	1.72	696.	-6.
MAR	54.2	21.74	14.39	1.91	599.	0.
APR	71.8	19.98	7.88	1.85	328.	7.
MAY	95.9	21.66	3.96	1.91	165.	13.
JUN	100.0	22.95	.96	1.85	40.	19.
JUL	100.0	24.54	.19	1.91	8.	21.
AUG	100.0	23.41	.52	1.91	22.	20.
SEP	100.0	22.79	2.31	1.85	96.	15.
OCT	81.3	20.36	6.32	1.91	263.	10.
NOV	33.2	12.92	12.12	1.85	505.	1.
DEC	26.3	13.96	17.81	1.91	742.	-5.
YR	49.1	235.06	103.07	22.48	4294.	

****ECONOMIC ANALYSIS****
OPTIMIZED COLLECTOR AREA = 39. M2
INITIAL COST OF SOLAR SYSTEM = $ 4900.
THE ANNUAL MORTGAGE PAYMENT FOR 20 YEARS = $ 449.

YR	INTRST PAID	END OF YR PRINC	DEPRC DEDUCT	PROP TAX PAID	INC TAX SAVED	BACKUP FUEL COST	INSUR, MAINT COST	COST WITH SOLAR	SAVNGS WITH SOLAR	PW OF SOLAR SAVNGS
1	352	4313	0	98	157	383	49	1312	-558	-553
2	345	4209	0	103	157	422	51	869	-41	-35
3	336	4097	0	110	156	464	55	922	-10	-8
4	327	3975	0	116	155	510	58	979	23	17
5	318	3844	0	123	154	561	61	1041	61	41
6	307	3703	0	131	153	617	65	1110	102	64
7	296	3550	0	139	152	679	69	1184	149	87
8	284	3384	0	147	150	747	73	1266	201	108
9	270	3206	0	156	149	822	78	1356	258	129
10	256	3013	0	165	147	904	82	1454	321	149
11	241	2805	0	175	145	995	87	1561	392	168
12	224	2581	0	186	143	1094	93	1679	470	186
13	206	2338	0	197	141	1204	98	1807	556	204
14	187	2076	0	209	138	1324	104	1948	652	222
15	166	1793	0	221	135	1456	110	2102	757	238
16	143	1487	0	234	132	1602	117	2271	875	255
17	119	1157	0	248	128	1762	124	2456	1004	271
18	92	800	0	263	124	1939	131	2659	1148	287
19	64	415	0	279	120	2133	139	2881	1306	302
20	33	0	0	296	115	2346	148	3124	1482	318

THE DISCOUNTED RATE OF RETURN ON THE SOLAR INVESTMENT(%)= 22.3
THE DISCOUNTED PAYBACK PERIOD IS(YR) 11.
YRS UNTIL CUMULATIVE SAVINGS=MORTGAGE PRINCIPLE 14.
PRESENT WORTH OF YEARLY TOTAL COSTS WITH SOLAR = $ 14244.
PRESENT WORTH OF YEARLY TOTAL COSTS W/O SOLAR = $ 16699.
PRESENT WORTH OF CUMULATIVE SOLAR SAVINGS = $ 2456.

TYPE IN CODE NUMBER AND NEW VALUE

BIBLIOGRAPHY

ASHRAE GUIDE AND DATA BOOK, HANDBOOK OF FUNDAMENTALS, American Society of Heating, Refrigerating and Air Conditioning Engineers, New York (1972)

ASHRAE GUIDE AND DATA BOOK, SYSTEMS, American Society of Heating, Refrigerating and Air Conditioning Engineers, New York (1973)

Beck E.J. and Field R.L., Solar Heating of Buildings and Domestic Hot Water, Technical Report R835, Naval Facilities Engineering Command (1976)

Beckman, W.A., Duffie, J.A., and Klein, S.A., "Simulation of Solar Heating Systems," Chapter 9 of the ASHRAE book, Applications of Solar Energy for Heating and Cooling a Building, ASHRAE GRP 170, American Society of Heating, Refrigerating, and Air Conditioning Engineers, New York (1977)

Bennett, I., "Monthly Maps of Mean Daily Insolation for the United States," Solar Energy, Vol. 9, 145 (1965)

Bliss, R.W., Jr., "The Derivation of Several 'Plate Efficiency Factors' Useful in the Design of Flat-Plate Solar Collectors," Solar Energy, Vol. 3, 55 (1959)

Carrier Corporation, System Design Manual, Vol. I, Load Estimation, Syracuse, New York (1972)

de Winter, F., "Heat Exchanger Penalties in the Double Loop Solar Water Heating Systems," Solar Energy, Vol. 17, 335, (1975)

Duffie, J.A. and Beckman, W.A., Solar Energy Thermal Processes, Wiley Interscience, New York (1974)

Engebretson, C.D., "The Use of Solar Energy for Space Heating: MIT Solar House IV," Proceedings of the UN Conference On New Sources of Energy, Vol. 5, UN, New York, 159 (1964)

Hesselschwert, A.L., "Performance of the MIT Solar House," Proceedings of the Symposium on Space Heating with Solar Energy, Massachusetts Institute of Technology (1950)

Hill, J.E., Streed, E.R., Kelly, G.E., Geist, J.E., and Kusada, T., Development of Proposed Standards for Testing Solar Collectors and Thermal Storage Devices, Technical Note 899, National Bureau of Standards, Washington D.C. (1976)

Hise, E.C. and Holman, A.S., Heat Balance and Efficiency Measurements of Central Forced-Air Residential Gas Furnaces, Report ORNL-NSF-EP-88, Oak Ridge National Laboratory, Oak Ridge, Tennessee (1975)

Hottel, H.C. and Woertz, B.B., "Performance of Flat-Plate Solar Heat Collectors," Trans. ASME, Vol. 64, 91 (1942)

Hottel, H.C. and Whillier, A., "Evaluation of Flat-Plate Collector Performance," Trans. of the Conference on the Use of Solar Energy, Vol II, Thermal Processes, 74, University of Arizona, (1955)

Jennings, B.H., Environmental Engineering, International Textbook Company (1970)

Klein, S.A. et al., TRNSYS - A Transient System Simulation Program, User's Manual, Report #38, Engineering Experiment Station, University of Wisconsin - Madison (1973)

Klein, S.A., "A Design Procedure for Solar Heating Systems," Ph.D Thesis, University of Wisconsin - Madison (1976)

Klein, S.A., Beckman, W.A., and Duffie, J.A., "A Design Procedure for Solar Heating Systems," Solar Energy, Vol. 18, 113 (1976)

Klein, S.A., Beckman, W.A., and Duffie, J.A., "A Design Procedure for Solar Air Heating Systems," Proceedings of the American Section Meeting of

ISES, Vol. 4, Winnipeg, Canada, August (1976), (to be published in Solar Energy (1977))

Klein, S.A., "Calculation of Monthly Average Insolation on Tilted Surfaces," Proceedings of the American Section Meeting of ISES, Vol. 1, August (1976) (to be published in Solar Energy (1977))

Klein, S.A., Beckman, W.A., and Duffie, J.A., Monthly Average Solar Radiation on Inclined Surfaces for 171 North American Cities, Report No. 44, Engineering Experiment Station, University of Wisconsin - Madison (1977)

Liu, B.Y.H. and Jordan, R.C.,"The Interrelationship and Characteristic Distribution of Direct, Diffuse, and Total Solar Radiation, Solar Energy, Vol. 4, 1, (No. 3, 1960)

Liu, B.Y.H. and Jordan, R.C., "Daily Insolation on Surfaces Tilted Toward the Equator," Trans. ASHRAE, 526 (1962)

Lof, G.O.G., Duffie J.A., and Smith, C.O., World Distribution of Solar Radiation, Report No. 21, Engineering Experiment Station, University of Wisconsin - Madison (1966)

Lof, G.O.G., El-Wakil, M., and Chiou, J.P., "Design and Performance of Domestic Heating Systems Employing Solar Heated Air - the Colorado Solar House," Proceedings of the UN Conference on New Sources of Energy, Vol. 5, UN, New York, 185 (1964)

Lof, G.O.G. and Tybout, R.A., "Cost of House Heating with Solar Energy," Solar Energy, Vol. 19, 253 (1973)

National Association of Home Builders Research Foundation, Insulation Manual for Homes and Apartments, Rockville, Maryland (1971)

Pettit, R.B. and Sowell, R.P., "Solar Absorptance and Emittance Properties," Journal of Vacuum Science Technology, Vol. 13, (1976)

Ruegg, R.T., Solar Heating and Cooling in Buildings: Methods of Economic Evaluation, U.S. Dept. of Commerce, NBSIR 75-712, July (1975)

Shurcliff, W.A., Solar Heated Buildings - A Brief Survey, 13th edition, Cambridge, Mass. (1977)

U.S. Dept. of Commerce, Climactic Atlas of the United States, Environmental Data Service, Reprinted by the National Oceanic and Atmospheric Administration (1974)

U.S. Dept. of Commerce, Monthly Normal of Temperature, Precipitation, and Heating and Cooling Degree-Days (1941-1970), National Oceanic and Atmospheric Administration, Climatography of the United States No. 81 (by state)

Ward, J.C. and Lof, G.O.G., "Long-term (18 years) Performance of a Solar Heating System," Solar Energy, Vol. 18, 301 (1976)

NOMENCLATURE

A collector area [m^2, ft^2]

C_F price of fuel [$/GJ, $/MMBTU]

C_{min} minimum capacitance rate in a heat exchanger [W/C, BTU/F]

C_p specific heat [J/kg-C, BTU/lbm-F]

d discount rate

DD monthly degree-days [C-day, F-day]

E total auxiliary energy [J, BTU]

f fraction of monthly load supplied by solar energy

$\mathcal{F}$ fraction of the annual load supplied by solar

F_R collector heat removal efficiency factor

F_R' collector-heat exchanger efficiency factor

G collector mass flowrate per unit collector area [kg/s, lbm/hr]

H_F heating value of fuel [J/unit]

$\bar{H}$ monthly average daily total radiation on a horizontal surface per unit area [J/month-m^2, BTU/month-ft^2]

$\bar{H}_d$ monthly average daily diffuse radiation on a horizontal surface per unit area [J/month-m^2, BTU/month-ft^2]

$\bar{H}_T$ monthly average daily total radiation on a tilted surface per unit area [J/month-m^2, BTU/month-ft^2]

i inflation rate [%/100]

I_T	rate of solar radiation incident on a tilted surface per unit area [W/m^2, $BTU/hr-ft^2$]
K	factor defined in equation 2.5
$\bar{K}_T$	ratio of monthly average actual to extra-terrestrial solar radiation
L	monthly total space and/or water heating load [J/month, BTU/month]
L_s	monthly space heating load [J/month, BTU/month]
L_w	monthly water heating load [J/month, BTU/month]
m	collector air flowrate per unit collector area [$l/s-m^2$, cfm/ft^2]
M	storage capacity per square meter of collector area [l/m^2, gal/ft^2]
n	day of the year, Jan 1=1, Dec 31=365
N	number of days in a month or number of years
N_F	unit of fuel
Q_T	total energy collected during a month [J, BTU]
Q_u	rate of useful energy collection [W, BTU/hr]
$\bar{R}$	ratio of monthly average daily total radiation on a tilted surface to that on a horizontal surface
$\bar{R}_b$	ratio of monthly average daily beam radiation on a tilted surface to that on a horizontal surface
s	angle between the plane of the collector and horizontal [°]
T_a	ambient temperature [C, F]
$\bar{T}_a$	monthly average ambient temperature [C, F]
T_{avg}	average collector fluid temperature[C, F]

T_c temperature of a cold stream entering a heat exchanger [C, F]

T_h temperature of a hot stream entering a heat exchanger [C, F]

T_i temperature of fluid entering a solar collector [C, F]

T_m temperature of mains supply water [C, F]

T_o temperature of fluid leaving a solar collector [C, F]

T_{ref} reference temperature [100 C, 212 F]

T_w acceptable temperature for domestic hot water [C, F]

UA building overall energy loss coefficient-area product [W/C, BTU/hr-F]

U_L collector energy loss coefficient [W/C, BTU/hr-F]

V volume of the packed bed per unit collector area [m^3/m^2, ft^3/ft^2]
daily volume of domestic water [l]

X dimensionless quanity defined by equation 5.3

X_c corrected value of X

Y dimensionless quanity defined by equation 5.4

Y_c corrected value of Y

α absorptance of the collector plate surface

α_n normal incidence absorptance

δ solar declination [°]

Δt number of seconds in a month [s]

ΔU change in internal energy of storage in month [J, BTU]

ε heat exchanger effectiveness

ε_c effectiveness of the collector-tank heat exchanger

ε_L effectiveness of the load heat exchanger

η collector efficiency

η_F furnace efficiency

$\overline{\theta}_b$ average incidence angle for beam radiation [°]

ρ ground reflectance

τ transmittance of the collector cover system

τ_n normal incidence transmittance

$(\tau\alpha)_n$ transmittance-absorptance product for radiation at normal incidence

$(\overline{\tau\alpha})$ monthly average transmittance-absorptance product

$(\overline{\tau\alpha})_b$ monthly average transmittance-absorptance for beam radiation

$(\overline{\tau\alpha})_d$ monthly average transmittance-absorptance for diffuse radiation

$(\overline{\tau\alpha})_r$ monthly average transmittance-absorptance for reflected radiation

ϕ latitude [°]

ω_s sunset hour angle for horizontal surface [°]

ω_s' sunset hour angle for tilted surfaces [°]